THE VERY EFFICIENT CARPENTER

THE VERY EFFICIENT CARPENTER

Basic framing for residential construction

Larry Haun

The Taunton Press

Cover photos: Larry Hammerness

The Taunton Press
Inspiration for hands-on living®

Printed in the United States of America
20 19 18 17 16 15 14

For Pros / By Pros®: The Very Efficient Carpenter
was originally published in 1992 by The Taunton
Press, Inc.

For Pros / By Pros® is a trademark of The Taunton
Press, Inc., registered in the U.S. Patent and Trademark
Office.

The Taunton Press, Inc., 63 South Main Street,
Newtown, CT 06470
e-mail: tp@taunton.com

Library of Congress Cataloging-in-Publication Data

Haun, Larry.
    For Pros / By Pros®: The very efficient carpenter :
  basic framing for residential construction /
  Larry Haun.
      p.   cm.
    Includes index.
    ISBN 1-56158-326-X
    1. House framing. I. Title.
  TH2301.H38   1998   92-12644
  694'.2—dc20   CIP

ACKNOWLEDGMENTS

As a carpenter and a teacher of carpenters, I have long seen the need for a different approach to teaching frame carpentry. That's why I decided to write this book and to make the companion videos. The list of those who helped along the way is long. Here's thanks to those who gave more than a little help.

The many people at The Taunton Press who gave me enthusiastic encouragement.

Jeff Beneke, my editor, who asked a million questions and shortened, clarified and added to the quality of the text.

My wife, Mila, and daughters Ninay and Risa, who helped with duplicating, computer problems and love.

My brothers Jim and Joe and my sisters Margaret and Loretta, for technical help and encouragement.

Bill Mauger, who taught me about video and how to work before a camera.

Jim Hall and his video crew, Jeff Fay and Tim Olson. Hard workers, all.

Roger Turk and my son Eric, for their photographic work.

Creighton Blenkhorn, director of UBC apprenticeship training in southern California, who got me in front of the camera to make training videos for apprentices.

Kelly Adachi, who taught me computer basics.

Joel Foss, OSHA, for information on safety.

Ed Franken, who helped me solve some framing problems.

John Gibson, for the house plans.

My nephew Larry Haun, who saved me when my computer went in unknown directions.

Marty Hittleman, for help with math.

Milt Rosenberg, building inspector, for technical help.

Jane Tokunow, for reading and reviewing materials.

This book is dedicated to carpenters and framers everywhere who love
the smell and feel of wood and enjoy building houses
that people can turn into homes.

CONTENTS

INTRODUCTION

In rural western Nebraska in the 1930s, no houses were being built, and I took the existing ones for granted, much as I did the trees, the hills and the constant wind. They just were. The only type of construction I knew was when my father set some fence posts in the ground, wrapped the perimeter with hog wire both inside and out, and stuffed the middle full of straw to form a wall and give some protection to a milkcow. Only when the high school was remodeled and added to did I begin to realize that buildings had to be brought into existence, that buildings have authors like books. This was a fascinating revelation, one that fascinates me to this day as I see new structures rise up out of the ground.

The carpenters who did that remodeling job came from another town, "over on the river," and wore white overalls with a lot of extra pockets. They carried big toolkits with shiny handsaws, levels, planes, squares, braces and bits, plumb bobs, chisels sharp enough to shave with, little hammers, big hammers, sledgehammers. I especially loved to watch the long curls of wood rising up out of a plane as it was pushed over a surface. I hung around so much that they finally put me to work as a waterboy and general gofer—at 50 cents a day. That summer made a lifelong mark on me.

We had some tools at home, of course: a small handsaw with teeth more rounded than pointed, a claw hammer with one claw missing, an old nicked wood chisel. There were no power tools because electricity wasn't available in our area. With these basic tools my father kept the animals sheltered, generally using nails salvaged from the sites of burned-down buildings. For my part, I was busy making toy propellers and kite struts from the lids of bushel baskets and the lovely soft wood of orange crates. I remember struggling for hours on end to cut off a piece of wood with the old handsaw. My best tool was a jackknife that I managed to keep somewhat sharp by borrowing a whetstone from a friend. So I was deeply impressed by the beauty and power of the tools those carpenters had. And the smell of their work area as they worked to shape the different woods is still with me. I have smelled it a thousand times since, but that first time was my memory marker.

In 1947, when I was 16, I helped an old carpenter build a house. He was a gentle old man, teaching an apprentice the basics of the trade. With me working in the summer and part time during the school year, it took us almost a year to nail it all together. His tools were sacred, cared for like fine jewelry, carefully wrapped and protected from rust, used with the utmost care and precision. He took great pride in his work, with good reason, for he was as much an artist as a craftsman.

But the postwar world was changing rapidly and leaving him behind.

Out of high school, I needed money for college, so when my brother invited me down to Albuquerque to help him frame houses, I jumped at the chance. Besides needing money, I was ready for a warmer climate. The great postwar housing boom was beginning with lots being cleared and foundations being poured all over town, and any willing worker had a job. We were still in white overalls, using only hand tools. As carpenters, we were expected to cut framing lumber with a handsaw, pour foundations, shingle roofs, lath, lay and finish hardwood floors and build cabinets. But as the need for housing increased, this began to change. The age of specialization was rapidly approaching. Builders started trying to apply to construction the mass-production, assembly-line techniques that Henry Ford used to build cars.

This was not a matter of decreasing the quality or durability of houses. Building codes became stricter year after year, as they do to this day. Even tougher was the Federal Housing Administration, which separately inspected all the millions of G.I. houses it financed. The old saying, "They just don't build houses the way they used to," is true. For the most part, they build them much better. But the great discovery, not yet fully understood or accepted, is that quality does not have to be sacrificed to speed and efficiency. In reality, it can be enhanced by them. One of the purposes of this book and the three companion videos is to persuade you of that fact.

In 1950 my brother and I moved on to Los Angeles, where I entered UCLA and joined the union as a journeyman carpenter. The following year my brother bought one of those G.I. houses in the San Fernando Valley — $400 to move in and monthly payments of $63, which included taxes and insurance. At last carpenters could afford to buy the houses they built — the American dream fulfilled. The demand, understandably enough, was enormous, and one of the great experiments in American ingenuity was evolving to meet it.

Carpenter wages then were under $2 an hour, but we found a builder in a Los Angeles suburb who was willing to let us do his framing for a flat fee per house. For a 900-sq ft., two-bedroom, one-bath house on a slab with a hip or gable roof, he paid us $90. A three-bedroom, 1,100-sq. ft. version went for $120. Both deals included jambs and window frames and a bit of siding, all complete and ready for roofing and plaster. My younger brother came out from Nebraska and joined us, and soon the three of us were framing one of these houses every day, more than doubling our previous wages. How we were able to do this is the subject of this book and the companion videos.

White bib overalls, which restricted movement, were replaced by pants and nail bags on belts. Hand tools gave way to an expanding selection of power tools. The hand tools that remained were changed. The traditional 16-oz. curved-claw hammer, for example, was replaced by a heavier straight-claw model with a big serrated striking face, capable of driving a 16d nail with one lick. More significant, perhaps, was that the old procedures changed as we sought ways to save a minute here and five there. Thousands of other carpenters were doing likewise, and we learned from each other. New tools were improvised; if they worked out, they were soon manufactured. The job of framing was broken down into subspecialties. Roof cutters, stairbuilders, sheathers, wall framers, detailers and many others began to refine and polish their own skills and techniques.

Where carpenters had traditionally been taught to "measure twice, cut once," we learned to measure by eyeballing, a much quicker method. We tried to group our tasks, finishing all of one job before moving on to the next. We developed new terminology, which you will become familiar with in this book: scattering, plating, detailing, stacking, and so on.

There was resistance to all this, of course. Some people just don't like change, others were afraid of losing their jobs. But the trend was too powerful, and the change went on — it still does, not only in carpentry but in most other trades as well. We went on to form a carpentry subcontracting company specializing in framing, and we soon found ourselves doing mainly multiple-unit apartment buildings, filling in the slow times with houses and commercial buildings. Our record times, which are not in the Guinness book, are a 34-unit building framed in four days and a 100-unit building framed in 13 days. The 100-unit building went from ground-breaking to tenancy in three months. It takes some luck, a lot of synchronization of trades, perfect on-time deliveries and a sharp and eager builder, but it can be done.

When I began writing articles for *Fine Homebuilding* magazine, I was curious why the folks at The Taunton Press would be interested in the tools and techniques of production framing. I knew that there is a bias against these methods of work in some parts of the country. They informed me that they believe, as I certainly do, that these methods can be used even on the finest custom homes, saving time and money without sacrificing quality. "Save where you can, spend where you must." That's the message of this book-video set.

GETTING STARTED

Tools

Plans, Codes and Permits

Lumber and Materials

1

TOOLS

Eric Haun

For many years in this country the craft and the tools of carpentry changed very little. Incredible structures, many of them works of art, were crafted with simple hand tools. Much of the Empire State Building, for example, was measured out with folding wooden rules. Often these tools were works of art in themselves, razor sharp and rust free, cherished by their owners and passed down to apprentices learning the trade from the master craftsmen. There were few power tools; the portable power saws that began to appear were looked upon as rough, imprecise and dangerous. But after World War II the demand for housing became so enormous that both the methods and tools used in construction began to change. An earlier carpenter, working with the sharpest 12-point handsaw, might have spent half a day fitting the joint on a 12x12 beam. Now, with the proper power tools, this job can be done in a few minutes. The power tools of today were practically nonexistent 45 years ago. Now the reverse is true—it is quite possible to work on a job where not one carpenter has a handsaw, let alone a folding wooden rule!

Basic framing tools

Very few tools are really needed to frame the average house, but the one that has increased efficiency tenfold is the portable circular saw. This saw is the workhorse of the construction industry. Many framers prefer a heavy-duty model with a 7¼-in. blade, which is fairly lightweight, rugged and relatively inexpensive. West of the Mississippi the preference seems to be for the type that is driven by a worm gear, whereas in the East the direct drive (side-

Using a circular saw

No matter how careful you are with a circular saw, at some point it is going to bind in the wood and kick back on you. So make safety precautions part of your work ethic. Don't wedge the blade guard up! Some cuts may be easier to make with the guard out of the way, but this eliminates your protection in the event of a kickback. Support the wood well, stand to the side of the line of cut, release the trigger just before the end of the cut so the blade slows down and keep your mind on your work.

Frame carpenters do a lot of plunge cutting with their circular saws. This is not a difficult or dangerous technique if you learn to do it right. Set the front of the saw table on the material to be cut, pick the guard up slightly with one hand to expose the blade to the wood, and start the saw. As you start the cut, release the guard and grab the upper handle of the saw. Then carefully drop the saw into the wood.

Always use sharp blades. Cutting with a dull blade is dangerous, because a lot of pressure has to be exerted to get it to do what a sharp blade does with ease. It makes it hard to control the saw. Speed and efficiency also suffer. And when changing a sawblade, take a second to unplug the power source first.

Eric Haun

The circular saw is the workhorse of frame carpentry. This worm-drive model is equipped with a 'Skyhook,' which allows it to be hung securely on a joist or rafter.

winder) is more common. For either model, a cross-cut blade will handle most framing tasks. Carbide-tipped blades are great for those who want to cut down on frequent sharpening. Add a good 12-gauge extension cord and nearly all of your basic wood-cutting needs will be met. A handy accessory for the circular saw is a "Skyhook" (available from Pairis Enterprises, 2151 Maple Privado, Ontario, CA 91761). This device folds out and hooks over a joist or rafter to hold the saw safe and secure.

Pneumatic nailers still haven't totally replaced the hammer, and perhaps never will. Carpenters are always on the lookout for a better-balanced, good-feeling framing hammer, and there are plenty on the market to choose from. For example, one now available is a hybrid that evolved by cutting the straight claws from a hammer and welding them to the face and handle socket of a rigging ax (hatchet). It has good balance and a flattened handle that fits well in the hand.

Most framers seem to prefer a 20-oz. to 24-oz. straight-claw hammer with milled (serrated) face and a fairly long (16 in. to 18 in.) wooden handle. Bear in mind that driving nails well is mainly a matter of wrist action and not hammer size or handle length. In my experience, wood handles seem to absorb shock better than metal or fiberglass handles. The straight claw can be stuck into heavy lumber to move it around the job site, and the hammer becomes an extension of the arm, an extra hand for a framer. It allows carpenters to pry plates apart easily in preparation for wall framing, to pick up headers and move them into position and to pull beams around without having to pick them up. The serrated face makes it less likely that the hammer will

Hammers are available in a variety of sizes and weights. Most framers prefer the straight-claw version.

slip off the nailhead while working, yet it does have its disadvantages. It always hurts to hit your thumb with a hammer, but it seems to hurt a bit more when it leaves waffle marks.

Sometimes a wooden handle can become slick and difficult to grasp, so it helps to rough it up a bit with a wood rasp or rub it with a bar of common kitchen paraffin, which makes for a firmer grip. You can even rub a bit of pitch from the wood on your hand to keep a slippery handle secure. In really dry weather, no amount of steel wedges can keep a handle tight in a head; slipping your hammer after work into a bucket of water kept by the garage door will help to keep the handle from drying out and the head becoming loose.

Hammer handles do break from time to time. When buying a new one, make sure the wood grain runs parallel with the hammerhead. Grain running at a right angle to the head makes it easy to break the handle. Once the new handle is tightly fitted into the head with wedges, wrap the first few inches below the head with electrical tape to make it stronger yet.

Sometimes nails bend or are driven in the wrong place and have to be pulled. An easy way to break a hammer handle is to hook the nailhead with the claws and pull directly on the handle. A better way

to remove the nail is to let the claws grip its shank nearest the wood. Then lever the nail out by pushing the handle to the side. This action will remove the nail about 1 in. Repeat until the nail is completely removed.

The traditional white overalls with their many pockets were nice to wear, especially in the colder parts of the country, but they restricted ease of movement. The leather belt with nail bags hanging from it solved this problem. Framers prefer a wide belt to distribute the weight, and wear bags to the sides or rear so they can grab a handful of nails even though they are bent over. Many other tool holders can also be fitted to the belt, the most common being a hammer loop. Insert a key ring with a clip through a hole punched in the belt and you have a place to carry a small adjustable wrench for changing sawblades. When the bags are loaded down with nails, the belt can weigh more than 12 lb., not a tremendous weight but noticeable after eight hours of work. Several years ago some tired framer finally fitted a pair of heavy red suspenders to the belt and heaved a huge sigh of relief; part of the weight was now on his shoulders, and he wondered why he'd never thought of it before.

Some commonly used framing tools are measuring tapes, a chalkline and refill bottle, pencil and keel, an awl and a dryline.

The most commonly used measuring tape is the push-pull type with a retractable metal blade. Carpenters doing layout work prefer the 25-ft. model to shorter tapes. On occasion, you may need a 50-ft. or 100-ft. steel tape for longer distances.

One of the most useful tools in a framer's nail bags is the chalkbox, used everywhere a straight line needs to be marked; it also works well as a plumb bob. Chalkboxes used to be filled by scooping chalk into them with a teaspoon, but it wasn't long before someone appeared on the job with a plastic shampoo bottle filled with chalk, a precursor of today's squeeze bottle that fills the box through a nozzle. Many framers like to use red or blue cement coloring rather than chalk because it doesn't wash away as easily in the rain.

The basic marking tools are a pencil and a piece of crayon. The carpenter's pencil is flat with fairly hard lead. The crayon is large and is called "keel" by people in the trade. These two can be kept together by using a short piece of plastic tubing, fitting a pencil in one end and keel in the other.

A carpenter needs to carry around a lot of tools and nails. The best way to do this is with a wide, sturdy belt equipped with various tool holders and nail bags and supported by heavy-duty suspenders.

Handy hand tools

The traditional carpenter's toolbox has given way to the carpenter's tool bucket as a means of carrying tools. This is most often a cleaned-out plastic bucket that used to hold drywall joint compound. A bag is now available that fits around the bucket, with many pockets to hold individual tools (Portable Products, 5200 Quincy St., St. Paul, Minn. 55112).

The small, triangular rafter square (often called a "Speed Square," which is one of the models available) has mostly replaced the try square and even the larger framing square in roof cutting and other jobs. The framing square, equipped with stair gauges or buttons and used to lay out stairs and frame roofs, has a simple L shape: The 2-in. wide, 24-in. long part is called the blade; the tongue is 1½ in. wide and 16 in. long. The "Squangle" (Mayes Bros. Tool Mfg. Co., Box 1018, Johnson City, Tenn. 37601) is an adjustable square that is handy for marking rafter tails. The drywaller's T-square can speed up marking plywood sheets for cutting.

A 2-ft. level fits nicely in the tool bucket, while a longer one can be carried and protected in a length of plastic plumber's pipe. A scratch awl comes in handy to hold one end of the chalkline on a wood floor; a movable weight, such as lead in a coffee can or a metal bar with a handle, will hold the line on a slab. A dryline, or long nylon string, helps layout of long walls.

A nail-pulling bar, or "cat's paw," can be used to remove nails driven in the wrong place, while a pry-bar helps to move heavy walls into position. A small 4-lb. to 6-lb. sledgehammer is useful for laying tongue-and-groove plywood, and you'll need a large caulking gun to dispense construction adhesive.

Screwdrivers, both standard and Phillips, and a set of Allen wrenches are often needed to repair equipment. Tin snips cut through the metal bands that hold loads of lumber together, although hammer claws may be handier. Pliers and sidecutters can be used to cut wire. A ratchet wrench with sockets works well for tightening bolts. Sometimes a wood chisel is needed when working with beams.

You can spend a lot of money for a fancier bucket, but a recycled joint-compound pail makes a great tool carrier.

Eric Haun

A drywaller's T-square (top) is handy for marking plywood. Other squares are necessary for laying out stair stringers and roof rafters.

A pocket calculator that works in feet and inches can increase efficiency, and a small book of rafter tables is a good way to figure rafter lengths. Even a builder's level is needed on occasion to set up level points over long distances.

A utility knife or a pocketknife has dozens of uses, from sharpening a pencil to cutting rolls of building paper. Duct tape is to today's carpenter what bailing wire was to those of yesteryear. Workers use it to wrap hammer handles, patch electrical cords, protect new sawblades and mend ripped jeans. Modern life, and certainly modern construction, is a little easier to manage with duct tape.

Site-built tools

Site-built tools such as sawhorses, scaffolds, ramps and ladders must be properly constructed, with no half measures. Often these tools are needed only for a short time, as with staging used to help set a heavy ridge beam in place, but they should be put together with care. What goes up can come down—don't let it be on you! Some of the most common devices that carpenters build on the job are a plumb stick (see p. 117) and push stick (p. 119) to help plumb and line walls and rafter templates (pgs. 140-141, 164, 176) to help lay out roof rafters.

Specialized tools

Many specialized tools have been developed by and for framers to simplify their jobs and to help them work more efficiently. Among these is a bolt-hole marker used to mark drill holes for bolts on sills (see p. 78). Several different types are available. A channel marker is a T-shaped metal or wood device that is used to lay out the location of outside and inside corners on the plates (see p. 92). Scribing the location of studs on the plates in preparation for framing is made easy with a layout stick (see p. 95).

Handy power tools

A ½-in. drill is a common power tool on framing jobs. It is used to drill holes for bolting two pieces of wood together, for attaching a beam to a metal saddle at the end of a post column or for bolting sills to foundations. A good selection of drill bits, both for wood and metal, is useful. The drill-bit sizes most often used are ½ in., ⅝ in. and ¾ in. Get some with short shanks and others with long shanks so you can drill to any required depth. A ½-in. masonry bit is sometimes needed to install bolts in concrete. For smaller jobs, like drilling pilot holes in decking, a 3/16-in. drill bit works fine. Once the holes

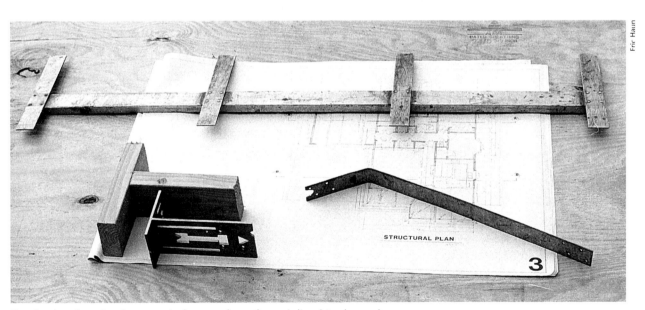

Production framing has created a number of specialized tools, such as the layout stick (top), corner and channel markers (left) and the bolt-hole marker (right).

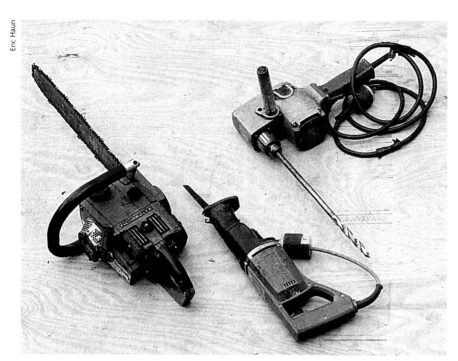

A chainsaw, a reciprocating saw and a drill are frequently needed by framers.

are drilled, an impact wrench with a set of sockets speeds up the process of tightening bolts.

Another handy tool, more often used on remodeling jobs, is the reciprocating saw equipped with both wood-cutting and metal-cutting blades. It pretty much replaces both the handsaw and the hacksaw. A power miter saw, or "chopsaw," can be used to cut blocks rapidly, but even quicker is the radial-arm saw, which can cut multiple pieces of wood. A beamsaw, a circular saw with a 16-in. blade, can be used for a variety of cutting jobs that are beyond the capacity of the 7¼-in. model. The gasoline-operated chainsaw also has its place in residential framing; on some jobs it is handy for cutting holes in roof sheathing for heater vents because you don't have to drag a power cord from place to place. A chainsaw attachment is available that turns a circular saw into a great tool for cutting beams and rafter tails (see pp. 145-146).

Table saws are seldom seen on rough framing jobs, but they can be used when available. For example, a table saw can speed up the process of ripping 2x stock for frieze blocks between rafters.

Powder-actuated tools are used to fasten wood and other materials to concrete and structural steel by means of a metal pin. They have a firing mechanism much like a gun and are available with either single or multiple power loads.

You don't need pneumatic nailers to build a house, but they can increase your production. The newer, lightweight models can handle different sizes of framing nails. Air compressors should be mounted on wheels for easy movement and need to have enough capacity for the number of nailers you will be using at any given time, each of which will also require sufficient lengths of air hose. There is even a palm-sized nailer that runs on air that can drive nails in those hard-to-reach places.

The most common source of power to run these electric tools is the temporary power pole, usually set up by a private company, inspected by the building department and then connected to the power grid by the electrical company. All this can take time, so remember to request an installation several weeks before your actual starting date. If the job is small, you may be able to buy power from a cooperative neighbor, or, if no power is available at all, there are several portable, gas-operated generators on the market, for sale and for rent, that will adequately fill your needs.

Nailers and air compressors have become standard equipment on framing jobs.

Another very useful, but expensive, tool is the forklift, which can be used to place loads of lumber exactly where you want them, saving many hours of hard work by human woodpackers. Some of these have masts that will go three stories high with booms that can set lumber loads in from the edge of the building. On larger framing jobs, cranes of various kinds are now often used, serving all trades, positioning loads anywhere they are needed on multiple-story buildings.

Safety equipment

Safety is a serious issue. There is too much involved not to make it part of your daily routine. Safe, protective clothing goes with the territory. Tank tops, shorts and sandals are nice and cool to wear, but a framing job calls for a little more protection for the body. Good shoes or boots, jeans and a long-sleeved shirt not only protect your body on a daily basis, but also lessen the risk of skin cancer in later years.

Safety devices should be as much a part of a framer's toolkit as a hammer. Whenever you or the people near you are hammering, sawing or using a nailer, you should be wearing protective eyeglasses. Those available today are lightweight and not uncomfortable to wear. The alternative isn't any fun to contemplate.

Ear protection comes in various shapes and sizes. Small, soft pieces of sponge are available that are easy to stuff into the ears, are not bulky or uncomfortable and yet give good protection from loud construction noises.

Leather gloves, hard hats and knee pads may not have to be worn all day, every day, but there are times when they can give you the protection you need. When you are doing a lot of sawing, especially in an enclosed place, wearing a good dust mask will help keep the sawdust out of your lungs. Meet an old carpenter with emphysema and let him tell you what he wished he had done when he was young. If you have to work around toxic fumes or sprays, then use a better-quality respirator.

Eric Haun

A good carpenter is a safe carpenter. Always protect your eyes, ears, head, feet, hands and lungs, and carry first-aid equipment so that injuries can be treated on the spot.

Use your head

Anyone who has worked around construction knows from experience that it can be dangerous. People do get hurt, and accidents don't always happen to "someone else." Most construction sites can be safe places to work if you make them so, and it's not just a matter of luck. It's important that you keep your work area clean and all tools and equipment sharp and in good repair, but safety is more than just brooms and new saws. Working safely has a lot to do with good training, attitude and presence of mind, meaning that you are actively concerned about your well-being and the well-being of those working with you.

They say that drinking and driving don't mix. Well, neither do drinking, or doing drugs, and driving nails, especially if you are working two or three stories off the ground. Slips and falls from work areas, especially from high places and ladders, account for more than one-fourth of all construction injuries. Work with a clear head and pay attention to what you, and those around you, are doing. Most jobs are only as safe as the person performing them.

Stay cool! Framing on a plywood deck in the hot sun can rapidly raise the body temperature. It is fairly easy for the body to run out of water on a hot day, leading to hyperthermia and heat exhaustion. Avoid overheating by drinking a lot of water.

Stay aware! Turning up a radio to full volume or wearing one with headphones can distract you at a time when your full attention is needed. And be doubly careful as the workday progresses. Injuries seem to happen more often as you tire later in the day.

It makes good sense to carry a well-equipped first-aid kit in your car or pickup. It doesn't have to be a traveling pharmacy, but should contain basic items such as Band-Aids and bandages, including a stretch bandage, sterile pads and cotton, a roll of gauze and adhesive tape, scissors and tweezers, an instant cold pack, a tourniquet and splint, iodine, sterile antiseptic wipes, antibiotic cream, an ammonia inhalant and aspirin. It's also smart to take a first-aid course from your local Red Cross chapter so you will know what to do in case someone gets injured.

Improper scaffolding is a particular problem on framing jobs and is the source of many injuries. Framers seldom need scaffolding except for a few minutes to raise a beam, sheathe a high wall or nail an otherwise unreachable spot. The temptation is to spike a block to the wall, nail on a crosspiece with a leg under it, throw on a plank and hop on. Most of the time this is cheap, fast and dangerous. Just because you work fast doesn't mean you can't work safely. People are more important than profits. In the final analysis, production should always take a back seat to safety.

PLANS, CODES AND PERMITS

Eric Haun

A new house comes to life because someone had a plan. Houses don't just happen. For carpenters, the plans often come from a builder and an architect, who combined their ideas and imagination on paper, joining concrete, steel, wood and other materials into a harmonious whole. Plans may come from an owner who bought a set of plans from a catalog or sketched his or her own. Wherever they come from, plans are the basic means of communication among architects, builders, carpenters, concrete contractors, electricians, plumbers and a host of other tradespeople. Learning how to read and interpret these drawings is an important step a carpenter must make to be able to turn ideas into reality.

Building plans are like road maps. If you want to drive from Florida to Alaska, for example, you look at a map knowing you can't visualize everything you will see along the way but that if you follow the lines and symbols properly you will arrive at your destination. The problem is that not everyone finds a road map easy to read, and house plans seem even more confusing. How does one go from visualizing an outline on a flat piece of paper to building a house that can be lived in?

Some people have the ability to look at a floor plan and visualize in their mind the walls, the doors, the roof, the entire building. Others find it impossible to see anything other than lines on paper. Fortunately, you don't have to be able to visualize the whole house in advance to be able to build it. Experience is a good teacher. Once you have built one house and seen how it goes together, you will find that the next time you look at a set of plans you

From Plans...to Real Walls

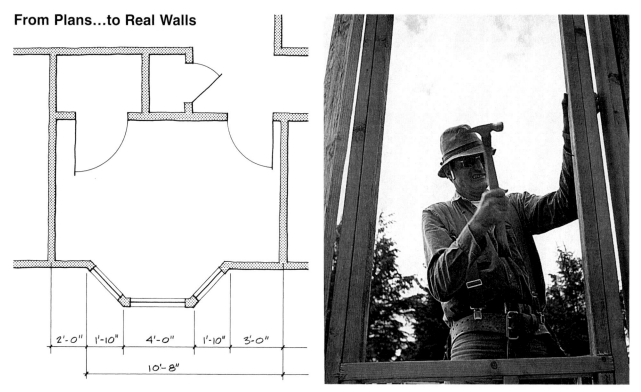

Carpenters must be able to bring an architect's floor plans to life.

Marking the Plans

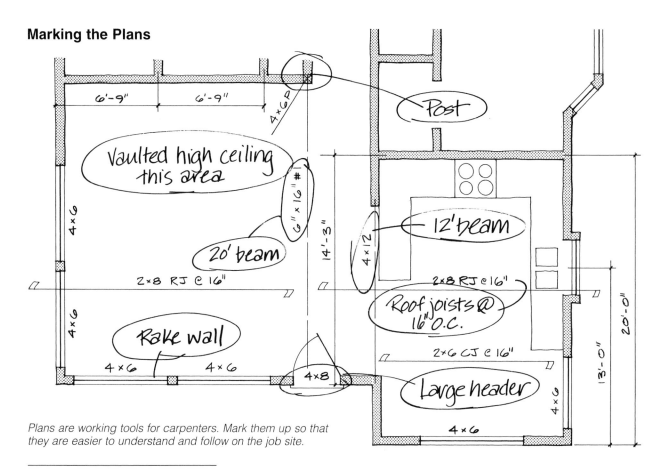

Plans are working tools for carpenters. Mark them up so that they are easier to understand and follow on the job site.

Reading Plans: Common Symbols and Abbreviations

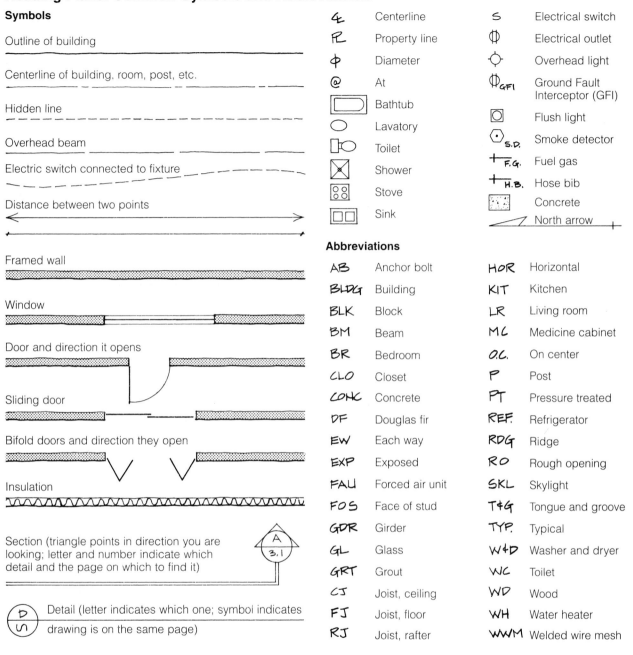

Symbols

Outline of building

Centerline of building, room, post, etc.

Hidden line

Overhead beam

Electric switch connected to fixture

Distance between two points

Framed wall

Window

Door and direction it opens

Sliding door

Bifold doors and direction they open

Insulation

Section (triangle points in direction you are looking; letter and number indicate which detail and the page on which to find it)

Detail (letter indicates which one; symbol indicates drawing is on the same page)

Symbol	Name	Symbol	Name
₵	Centerline	S	Electrical switch
P̶L	Property line	∅	Electrical outlet
φ	Diameter	○	Overhead light
@	At	∅GFI	Ground Fault Interceptor (GFI)
	Bathtub	▢	Flush light
	Lavatory	S.D.	Smoke detector
	Toilet	F.G.	Fuel gas
	Shower	H.B.	Hose bib
	Stove		Concrete
	Sink		North arrow

Abbreviations

Abbr.	Meaning	Abbr.	Meaning
AB	Anchor bolt	HOR	Horizontal
BLDG	Building	KIT	Kitchen
BLK	Block	LR	Living room
BM	Beam	MC	Medicine cabinet
BR	Bedroom	O.C.	On center
CLO	Closet	P	Post
CONC	Concrete	PT	Pressure treated
DF	Douglas fir	REF.	Refrigerator
EW	Each way	RDG	Ridge
EXP	Exposed	RO	Rough opening
FAU	Forced air unit	SKL	Skylight
FOS	Face of stud	T&G	Tongue and groove
GDR	Girder	TYP.	Typical
GL	Glass	W+D	Washer and dryer
GRT	Grout	WC	Toilet
CJ	Joist, ceiling	WD	Wood
FJ	Joist, floor	WH	Water heater
RJ	Joist, rafter	WWM	Welded wire mesh

will find it easier to visualize the finished house in three dimensions. With a little more experience, you may begin to relate to the standard, if overstated, joke among seasoned carpenters that the only reason they need a set of plans is to make sure they are building on the right lot.

Start by studying the plans at home. You will need to learn a whole new language of symbols and abbreviations; the chart above gives some of the more common ones. Orient yourself by finding out which direction is north and how the building sits on the lot. Use a marking pen to note the lengths and widths of lumber to be used and any unusual elements, as shown in the bottom drawing on the facing page.

Plan scales and dimensions

Plans (or "prints"—the term "blueprint" is no longer commonly used), like maps, are drawn to scale, and this scale is noted at the bottom of the page. The most common plan scale is ¼ in. to 1 ft., which means that each ¼ in. of line on a plan represents 1 ft. of the actual house. One inch, then, represents 4 ft. You must know the scale because not every dimension is given on every set of plans. An unmarked wall section, for example, may measure 1⅛ in. If the plans have a scale of ¼ in. to the foot, then the actual length of the wall will be 4 ft. 6 in.

Scales can be other than ¼ in. Sometimes the scale may vary on a single page, as with a detail drawing that has been blown up for greater clarity. Plans for large buildings may be drawn at ⅛-in. scale to keep them to a manageable size. Big mistakes can be made by assuming that the scale is always the same, so be sure to check it. A triangular architect's scale will come in handy when reading plans.

As shown in the drawing below, dimensions are given in several ways, depending on the custom of the architect. The full-size measurement of a building is usually given from outside to outside of the plates. The distance between an exterior and an interior wall is usually marked from the outside to the center. The distance between interior walls is fre-

quently marked from wall edge to wall edge, or center to center. The lines drawn by the architect should clearly indicate the starting and ending points of every dimension.

Types of plans

A normal set of house plans for carpenters consists of a plot (or site) plan, a foundation plan, a floor plan, a framing (or structural) plan and elevations. Additional components include sections, details, specifications and schedules.

The plot plan (see the top drawing on the facing page) gives an overall view from above, showing the shape and dimensions of the property and the size and location of the building to go on it. It might also include compass direction, contours, existing streets, utilities and sometimes even the location of trees. It is used mainly by the general contractor to make sure that the house is located properly on the lot.

Foundation footings, walls and piers are shown on the foundation plan (see the bottom drawing on the facing page). A concrete contractor, for example, will use this plan along with details (see p. 18) to learn the depth and width of footings. Framers often use the foundation plan to determine the elements of a wood floor system, such as the size, direction and spacing of girders and joists. If the house is going to be built over a basement, this is where to find the location and size of the stairway. Once the floor has been sheathed, carpenters turn their attention to the floor and framing plans.

The floor plan (see the drawing on p. 16) gives a bird's-eye view of a horizontal surface. A lot of information can be crammed onto this plan. The size and arrangement of all the rooms can be determined by a quick look at the floor plan. Sometimes the architect will draw separate pages detailing electrical, plumbing and heating, structural and other elements, but often much of that information will be found on the floor plan. If the house is going to be two or more stories, a separate floor plan will be drawn for each level. The floor plan may indicate the size of lumber needed for headers, posts and beams. The size and spacing of ceiling joists and roof rafters can be found here, and the direction they will run should be indicated by a line with an

Marking Dimensions

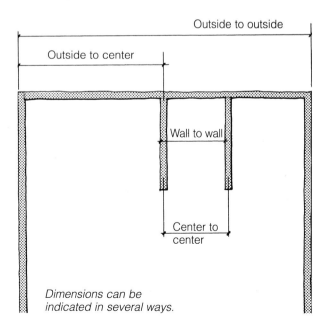

Outside to outside

Outside to center

Wall to wall

Center to center

Dimensions can be indicated in several ways.

Plot Plan

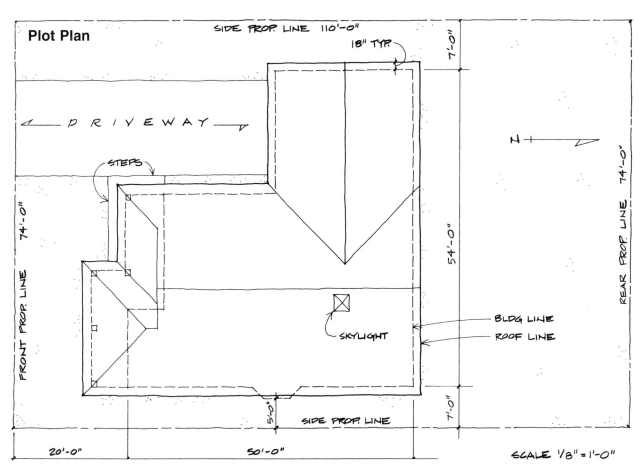

SIDE PROP. LINE 110'-0"

18" TYP.

7'-0"

← D R I V E W A Y →

STEPS

N

FRONT PROP. LINE 74'-0"

REAR PROP. LINE 74'-0"

54'-0"

SKYLIGHT

BLDG LINE
ROOF LINE

5'-0"

SIDE PROP. LINE

7'-0"

20'-0"   50'-0"

SCALE 1/8" = 1'-0"

Foundation Plan

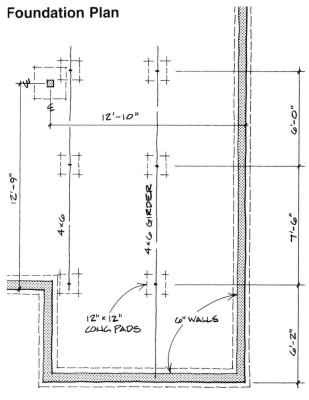

12'-10"

6'-0"

12'-9"

4×6

4×6 GIRDER

7'-6"

12" × 12"
LONG PADS

6" WALLS

6'-2"

The concrete contractor uses the foundation plan to form a base upon which to build the house.

Floor Plan

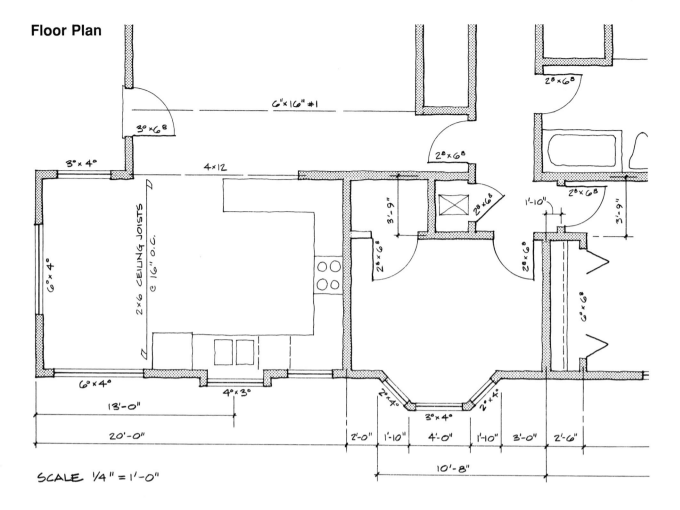

SCALE ¼" = 1'-0"

arrow. Any openings through the joists, such as for stairs, attic access or skylights, can be found on the floor plan.

A separate framing plan is one way to make floor plans less cumbersome. The framing plan can note window and door sizes, and sizes of headers, post-and-beam connectors and almost every other stick of wood that will be used in a building. It can indicate shear-wall location and any engineering or code requirements to help make the building better able to withstand earthquakes or high winds.

Some of the information you will need to build a house will require going to the "elevations." Elevations are side views. They are often labeled north, south, east and west (sometimes front, rear, right and left). They give complete vertical views of how the structure will appear, including the foundation, siding and trim, roof style and pitch, the length of overhang at the eaves, and so on. (Roofing infor-

mation is also sometimes shown on a separate roof plan.) A floor plan can show the location of windows and doors, while elevations can give their heights off the floor. Many radically different houses can be built to the same floor plan. Often those differences will be very apparent on the elevations. Interior elevations may also be given, particularly to show details of cabinets in baths and kitchens.

Section views give another perspective. Slice down through the house, just as you would through an apple, remove one half and stand back and look at the other. This is a section. It gives you a vertical view of the foundation, sills, floor joists, subfloor, and so on. A line through the house on a plan drawing will indicate with arrows from which direction the section drawing is being viewed. These lines may have reference numbers by them, indicating the plan page on which the section will be found and the drawing number on that page.

Elevation

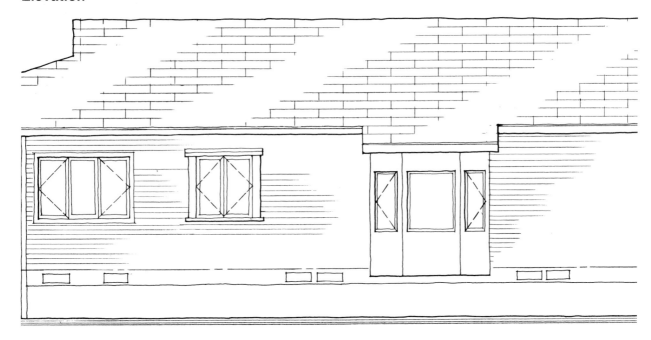

EAST ELEVATION

Section View

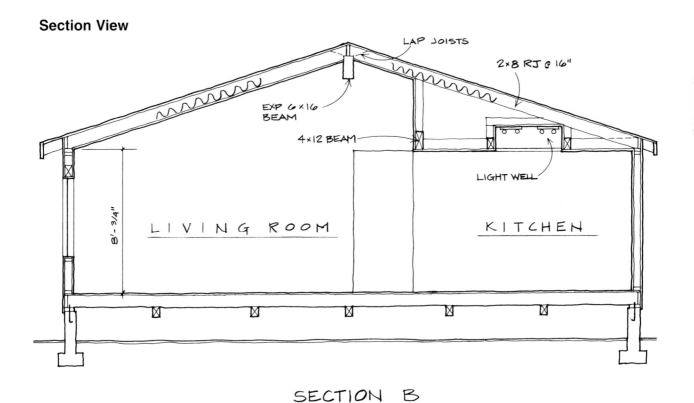

LAP JOISTS

2×8 RJ @ 16"

EXP 6×16 BEAM

4×12 BEAM

LIGHT WELL

8'- 3/4"

LIVING ROOM

KITCHEN

SECTION B

Detail Views

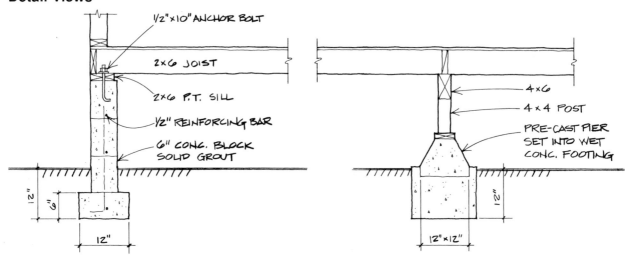

EXTERIOR FOOTING

- 1/2"x10" ANCHOR BOLT
- 2x6 JOIST
- 2x6 P.T. SILL
- 1/2" REINFORCING BAR
- 6" CONC. BLOCK SOLID GROUT
- 12"
- 6"
- 12"

PIER-GIRDER

- 4x6
- 4x4 POST
- PRE-CAST PIER SET INTO WET CONC. FOOTING
- 12"
- 12"x12"

Schedules

GENERAL NAILING SCHEDULE (PARTIAL)

POST TO PIER PAD · · · · 3-16D OR 4-8D
GIRDER TO POST · · · · 3-16D OR 4-8D
JOIST TO SILL OR GIRDER · 2-16D OR 3-8D
RIM JOIST TO JOISTS · · · 16D TOP & BOTT.
RIM JOIST TO SILL · · · · 16D @ 16" O.C.
STUD TO BOTT. PLATE · · · 3-16D OR 4-8D

SHEATHING

FLOOR PANELS TO JOISTS @ 16" O.C.
    8D - 6" O.C. OUTSIDE EDGES
       - 8" O.C. JOINTS
       - 12" O.C. FIELD

WALL PANELS TO STUDS @ 16" O.C.
    8D - 4" O.C. OUTSIDE EDGES
       - 6" O.C. JOINTS
       - 12" O.C. FIELD

ROOF PANELS TO RAFTERS @ 24" O.C.
    8D - 6" O.C. OUTSIDE EDGES
       - 8" O.C. JOINTS
       - 6" O.C. FIELD

Detail drawings give a close-up view of a small or complex feature of the house. They are either placed on a page all their own or scattered on the edges of the plans. Detail drawings are normally drawn to a larger scale to give enlarged views of such things as foundations, difficult stairways, complicated cornices and elaborate post-and-beam connections.

A set of house plans may include a specification (or spec) sheet, which spells out legal requirements, building codes, specific materials to be used, who is to do what work and exactly how the work will be done. Various schedules will also be included with most plans. The window and door schedules, for example, will indicate the quantity, size and style of windows and doors. The floor plan may indicate each door and window with a number or letter that is referenced to the schedule. The nailing schedule spells out the quantity, size and spacing of the nails needed to construct the building.

Codes and permits

Codes and permits, and the enforcers who go with them (i.e., inspectors), exist to protect home owners from unscrupulous and incompetent builders. They also exist to protect owner-builders and weekend do-it-yourselfers from themselves. To cut costs or to satisfy personal whims, people have been known to build houses that can't withstand a minor earthquake or a good gust of wind, that are fire traps because of amateur electrical wiring or that are health hazards because of shoddy materials or poor workmanship. Neighbors and future owners also deserve protection from such practices.

For these kinds of reasons building departments and building codes, though sometimes a nuisance and an expense, are necessary for all of us. They ensure that a new house will be built to proper standards with appropriate materials and have adequate lighting, ventilation, insulation, fire protection and security measures. They act as the quality control for the building industry.

Once a builder has a set of plans and a lot on which to build, the next step is to go to the building department and request the permits needed. Building departments in small communities may want to know only what you are planning to build, whereas those in larger cities may require a full-scale, detailed review of every aspect of the plans. Once the department is satisfied that construction will be up to their specifications, they approve the plans and issue a permit to build. Approved plans are kept available for inspections and should not be used as working plans. As actual construction proceeds, many building departments have inspectors who check, step by step, to see that the work is being done to code and in line with the approved plans. In such cases, for example, you wouldn't want to cover the walls of a framed house with drywall until an inspector had seen and approved the framing, wiring and plumbing.

There is no single code for all of the United States. There are national codes that more or less cover specific parts of the country (the Uniform Building Code in the West, the Standard Building Code in the South) and model codes that are used throughout the country (the National Building Code). But thousands of localities have adopted

Building Permit

their own specific codes that adapt the national models to their own situations. Houses being financed by the Veteran's Administration or the Federal Housing Administration generally must meet requirements above and beyond the local code. The Uniform Plumbing Code and the National Electrical Code guide those trades. Many communities, especially large cities, publish a shortened version of the code that gives a summary of the information framers need to know in order to build a typical one- or two-story wood-frame building.

When in doubt—and even when you're not in doubt—check with the building department or the local building inspector. They can keep you from making some very expensive mistakes. And always remember the First Rule of Construction: It is always cheaper to do it right the first time.

LUMBER AND MATERIALS

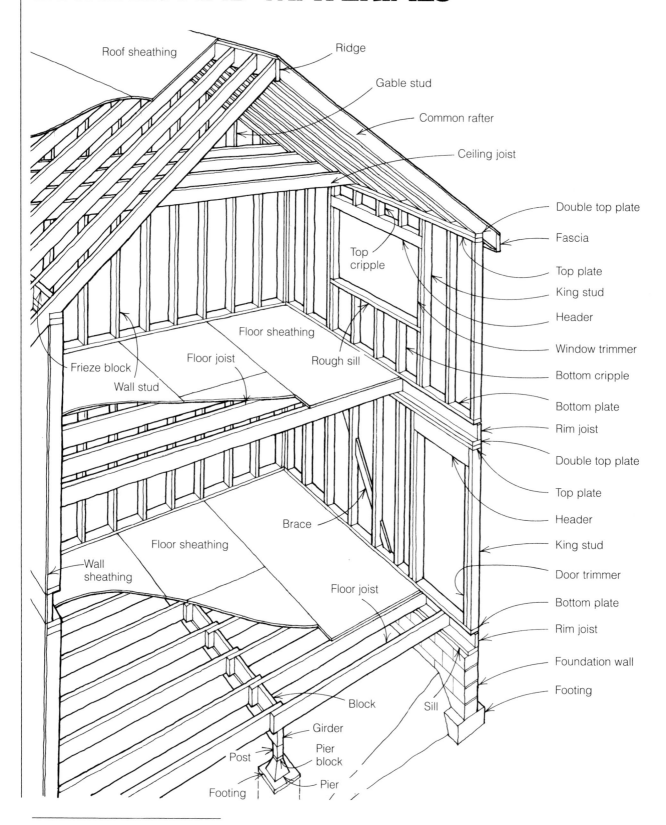

Roof sheathing

Ridge

Gable stud

Common rafter

Ceiling joist

Double top plate

Fascia

Top plate

King stud

Header

Window trimmer

Top cripple

Bottom cripple

Bottom plate

Rim joist

Double top plate

Top plate

Header

King stud

Door trimmer

Bottom plate

Rim joist

Foundation wall

Footing

Floor sheathing

Floor joist

Rough sill

Frieze block

Wall stud

Brace

Floor sheathing

Floor joist

Wall sheathing

Block

Sill

Girder

Pier block

Post

Pier

Footing

Estimating lumber and materials needs on most single-family residences is fairly simple, but it does take some experience to get it right. Learning how to build a house requires that you learn about the materials that go into it. No one wants to work with building materials that are unsafe, such as paint that has lead in it, asbestos-laced insulation or flawed wood. You need to know when and where to use different types of wood. Redwood is beautiful but brittle, and it might break if used for a ridge beam carrying a roof load. You need to know what size lumber to use. You wouldn't want to nail in 2x4 joists if they are not large enough to carry the load.

Materials to be used in a building project are governed to a great extent by local building codes (see p. 19). A framer doesn't have to know all of the codes, but must pay attention to what the plans specify when making up a lumber and materials list.

Once you have framed a house or two and learned how to read plans, you will be able to work up a lumber and materials list without much trouble. For framers, the list covers everything from the first sill on the foundation through the fascia board on the rafter tails, as well as all the hardware needed to put them together. The list is often the responsibility of the contractor, who can make it personally or submit the plans to a lumberyard and let them do it. Many lumberyards now use a computer to make a lumber list quickly. But making your own list gives you the opportunity to inspect the plans closely, to note areas that need special attention and generally to get acquainted with the house before you start. If you are buying the lumber, once you have a list, shop around for the best price.

Your list needs to spell out the type, size, quantity and grade of the materials. There is no need to supply board footage (the volume of lumber), since lumber companies can make the conversion if they need to. If you don't see lumber grades specified in the plans, check your building code; most spell out minimum grades required. Your lumber company may also have this information. For example, the most common framing lumber in the West is Douglas fir with a grade of #2 or better.

jim haun construction, inc
FRAMING CONTRACTORS
6846 VALJEAN AVENUE — VAN NUYS, CALIFORNIA 91406
780-1919

LUMBER LIST FOR __Peterson House__ ADDRESS __Coos Bay, Oreg.__ PHONE _____
JOB ADDRESS _____

LINEAL FEET	BOARD FEET	NO. OF PIECES	SIZE	LENGTH	TO BE USED FOR	GRADE
					All wood #2 grade or better	
					unless noted otherwise.	
		11	2X6	16'	Pressure-treated sill	
36			4X4		Girder posts	
		7	4X6	18'	Girders	
		8	5	12'		
		6	4X4	8'	Porch joists	
		10	5	14'		
		44	2X6	20'	sheathing — porch	cedar
		45	2X6	16'	Floor joists	
		42	5	18'	5	
		8	5	20'		
		33	¾ X4X8		T&G floor sheathing	OSB
52			2X4		P.T. garage sill	
500			2X6		Plate — outside walls	
1025			2X4		wall plates	
		160	2X6	92¼"	studs	
		340	2X4	5	5	
		6	2X6	14'	Rake-wall studs	
		15	1X6	12'	wall braces	
		12	3/8X4X8		Sheathing — rake wall	Struc1
		4	4X4	14'	Headers & posts	
		6	4X6	14'	5	
		1	4X8	4'		
		1	4X12	14'	Kitchen beam	
		1	4X16	18'	Garage header	
		1	6X16	22'	Ridge beam	#1
		7	4X4	8'	Porch posts	
		1	4X8	8'	Porch beams	
		1	5	14'	5	
		1	7	16'	5	
		1		22'	5	
		10	2X4	8'	Porch joists	
		10	5	14'	5	
		4	2X8	24'	Garage rafter ties	
		37	2X6	14'	Ceiling joists	
		5	5	16'	5	
		16	5	20'	5	
		58	2X6	20'	Rafters	
		16	2X8	20'	Cathedral — rafters	
		40	2X6	16'	Garage rafters	
		24	5	12'	Hip-roof rafters	
		4	2X8	14'	Hips & valleys	
		3	5	22'	Ridge	
		14	2X8	20'	Fascia	cedar
750			1X6		Shiplap starter board	cedar
					— small V-joint	
		62	½ X4X8		Roof sheathing	OSB
		HARDWARE				
		52	½"		Bolt nuts (anchor)	
		52	5		Bolt washers	
		4	50# 16d		Nails — sinkers	
		1	50# 8d			
		8	1/8 X 1½ X 18		metal plate straps	
		116			Hurricane ties	
		4	4X6		Joist hangers — U-type	
		22	2X6		5	

This Lumber List is supplied as a service only — No responsibility is assumed.

Lumber and Materials List

Manufactured lumber

As dimension lumber becomes less readily available, more manufactured lumber products are being used by builders. These include common items like plywood and oriented strand board (OSB) and newer products like wooden I-beam joists, such as TJI joists (Trus Joist Corp., P.O. Box 60, Boise, Idaho, 83707). These joists are made with a plywood web glued into flanges in a top and bottom chord (see the drawing on p. 42). They come in various depths and have knockouts in the web through which electrical conduit or water pipes can be run. Wooden I-joists are lightweight, easy to handle and can span up to 60 ft., making it possible to create very large rooms. Another advantage is that they are always straight, which means that the floors will be flat and level. Regular joists shrink faster than beams and can cause a floor to develop humps and squeaks.

A disadvantage of using wooden I-beam joists is that joist layout has to be done with more care. It is necessary to know the exact location of plumbing and heating runs, stairwells, and so on. These joists are engineered units, and any cuts or notches in them can seriously weaken their structural capacity. Companies that supply these joists also supply the hangers, hardware and information on installation, which is much like that of standard 2x joists.

Engineered beams, often made from laminated-veneer lumber (LVL), come in several widths and depths up to 60 ft. long. They are available in 3½-in. widths, which means they can be used for headers in a 2x4 wall; they can easily be furred out 2 in. more for a 2x6 wall. Beams and headers that are made from laminated and glued lumber have been available for years and won't twist and split like some species of solid-stock lumber.

Estimating tools

To begin making up a lumber list, you will need a couple of basic estimating tools and a set of plans (see pp. 14-18). A scale tape or an architect's scale will allow you to translate scaled dimensions on the plans to real dimensions. A pocket calculator is indispensable for keeping your math honest. The Scale Master (Calculated Industries, 22720 Savi Ranch Parkway, Yorba Linda, Calif. 92687) is a new tool that makes scaling plans even simpler. It is a hand-held digital gadget that quickly measures the length of any line no matter what the scale. The same company offers another useful tool, a pocket calculator that gives calculations in feet and inches, such as 10 ft. 6 in. instead of 10.5 ft.

Sills

Make the lumber list in exactly the order in which it will be used; the lumber company will put the first item on top of the load when they deliver. Most often on a house with a crawl space or basement, the first pieces of lumber you'll need are the sill plates. Specify on the list that this lumber must be pressure treated. Pressure-treated lumber is treated with substances that repel termites and inhibit dry rot, a fungus growth that can rapidly destroy wood. Working from the plans, measure the length of the foundation walls that will be covered by sills, add 5% extra, then divide by 16, which is the standard length of sills used by many builders, to determine the number of pieces needed.

Once your lumber list has been processed, the lumberyard will deliver the order in individual loads called lifts.

Posts and girders

The size of girders (often 4x6) and posts (often 4x4) will be given on the plans. Measure the lengths of all girders and then write the lengths with colored pen on the plans. Girders are often laid out with a post every 6 ft., so 12-ft. and 18-ft. lengths are common. Girders must be sized so that they break over a pier post so the ends can be properly supported. Writing the lengths on plans with a colored pen now will make installing them go faster when you are actually building.

If you are building over a basement, the supporting posts are usually 8 ft. long. Simply count the number you need and order them in 8-ft. or 16-ft. lengths. When building over a crawl space, each post will be around 16 in. long. Count the number needed, one for each pier, multiply by 16 in. and divide by 12 to get the number of lineal feet required. Post material can be ordered in random lengths.

Joists

Floor joists must be the proper length to span the building or lap on a girder. Check all lengths and mark them on the plans. If the joists are to be spaced 16 in. on center, the easiest way to estimate the number needed is to order one per foot. The extras will be used for rim joists, double joists under parallel walls, and blocks at the lap. Some builders, especially on larger jobs, will measure the lineal length of the rim joists and blocks and, as a cost-saver, order these from #3 lumber instead of #2 or better.

Floor sheathing

Floors are often sheathed with tongue-and-groove plywood or OSB. Determine the number of square feet in the floor by multiplying the length of the building by the width, subtracting or adding for any offsets. Divide this figure by 32, which is the number of square feet in a 4x8 sheet, and add 5%. Check the plans and/or code for thickness and grade.

Plates

Before calculating the lumber needed for wall plates, check the plans to see which walls are 2x6 and which 2x4. Each should be marked on the plans with a different colored pen to make it easier to work accurately. Then measure all the walls. If you are building on a wood floor (i.e., over a basement or crawl space), each wall will have three plates—one on the bottom and two on the top. So simply take your measurement of wall length and triple it. If you're building on a concrete slab, double the figure, keeping in mind that the bottom plate on a slab must be pressure treated. To be safe, you should add 10% to 15% to the figure you arrive at. You should not have to specify lengths for the plates. It is fairly standard practice for lumber companies to ship longer, random lengths for plate stock.

Studs

Estimating the number of studs needed is easy. For 16-in.-on-center walls, simply figure one stud for each lineal foot of wall. The extras will make up corners, partition intersections, trimmers and cripples. For example, if you need 350 ft. of sill stock, order 350 studs for walls. When studs are 24 in. on center, order one for every 2 ft. of wall and add 15% for extras. A standard stud length for many parts of the country is 92¼ in., and these studs can be delivered precut. Studs are cut this length because with three 1½-in. plates, the wall will be 96¾ in. high once it is framed. This leaves room for ½-in. drywall on the ceiling and 8-ft. high drywall on the walls.

Headers and cripples

Calculating the lumber needed for door and window headers takes a little more time and care. Each opening has a length and a header size listed on the plans. Order 4 ft. of stock for a 3-ft. opening, 5 ft. for a 4-ft. opening, and so on. Note by each length the size of the stock needed, like 4x4 or 4x6. Put a check on the plans at each opening as it is finished so you don't accidentally count it twice. Then add up the footage needed for each header size and order it in longer lengths. Longer headers, such as those used for a garage door, will of course require heavier stock. Some door headers may be in non-bearing walls and can simply be flat 2x4s, so no extra material needs to be ordered for them.

Some builders standardize the header size, at times making headers larger than required so that most of the top cripples can be cut the same length. Most window openings in the average house are un-

Lumber for Framing a Wall

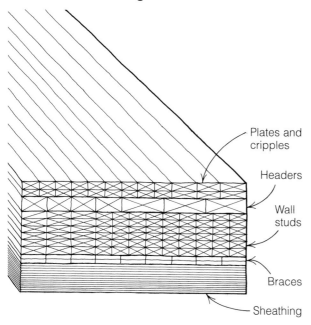

Plates and cripples

Headers

Wall studs

Braces

Sheathing

der 5 ft., for example; all the headers might be 4x6s even though a 4x4 would be adequate for some. If the headers are to be built up out of 2x stock, the length of the material required will be doubled, plus a sheet or two of ½-in. plywood, to be ripped to width and sandwiched between the 2xs so that the finished header will be 3½ in. wide. Headers in 2x6 walls can often be made with flat 4x6s, from 6x stock or from 4x stock with a 2-in. furring strip nailed on.

If cripples for doors and windows are to be cut on the job, plan to use scraps left over from joisting and plating and from bowed or twisted plate and stud stock. With a little effort, a house can be built with a minimum of waste. Some builders find it cost-effective, especially when building multiple units or a housing tract, to make up a header and cripple list (see p. 84) for the lumber company, which then does the cutting at the yard. Some companies do this for no charge as a loss leader; they can use cheaper grades, and older lumber they might otherwise have to discard or cut up for firewood.

Posts and beams

Check the length and size of any beams, such as those used as a ridge to carry rafters, and the posts they rest upon. Mark this information on the plans. Beams will sometimes be graded #1 or select structural or be made with laminated stock. Engineered beams may be required for particularly long spans or heavy loads, so check the details carefully.

Braces

The bracing that walls require depends on local code and on whether or not they will be sheathed. If you will be bracing all the walls with metal angle braces or wooden let-in braces, figure one at each end and one for every 25 ft. of wall. It is advisable to put a brace in every wall where it is possible to do so. Some walls will be too short or have too many openings to have room for a diagonal brace. If you are using wooden braces, it is best to order 12-ft. 1x6s, although 10-footers will work. Metal braces are available in 12-ft. lengths, but 10 ft. works well on standard 8-ft. walls.

Wall sheathing

If the walls are to be sheathed, the number of sheets needed for a one-story building is figured by taking the length of wall space to be covered and dividing by 4 ft., the width of each plywood panel. Pay no attention to door or window openings, except large ones such as garage doors or patio openings. There is usually no need to add 5% extra because the material cut out for doors and windows can be used to fill in above and below these openings, thus avoiding much waste. The standard length of plywood is 8 ft., but it can be purchased in longer lengths.

Ceiling joists and rafter ties

Ceiling joists can be figured somewhat like floor joists, although you don't need the extras for doubles, rims and blocks. When joists are spaced 16 in. on center, you can find the number needed by dividing the length of the building by four, then multiplying the result by three and adding one more for the end joist. When they are 24 in. on center, divide the length of the building by two and add one more; note and mark the different lengths and sizes needed to cover different sections of the building.

In the garage, rafter ties rather than ceiling joists are nailed in at 4 ft. on center to tie the building together. These frequently need to be from wider stock, like 2x8s, because even though they carry no weight they span a long distance. Measure in 4 ft. from the end wall and order one for every 4 ft. of wall. When rafters run at a right angle to the joists in a house, they too need to be tied across every 4 ft. to opposing rafters. Usually this rafter tie is made with 1x4 or 1x6 stock.

Rafters and other roof stock

If the roof is going to be framed with traditional rafters, the number of rafters needed can be calculated much like ceiling joists. On simple gable roofs with rafters spaced 16 in. on center, remember to double your figure to include rafters on both sides of the ridge. If the rafters are 24 in. on center, take the length of the building in feet and add two. This will give you enough stock for both sides of the ridge. If barge rafters are required, add two for each end. One way to determine rafter length is to measure the rafters, from the ridge to the fascia, on the elevation or roof plan. Hip and valley stock is usually 2 in. wider than the common rafters.

Order enough extra stock to cut a frieze block between every two rafters at the plate line, the full length of the building on two sides for a gable roof, all four sides for a hip. The ridge on a gable roof also will be the length of the building, though sometimes the plan will indicate that it runs beyond this point to catch and help carry barge rafters. Measure on the plans the ridge length you need for a hip roof. Note that the ridge stock must be 2 in. wider than the rafter stock (2x8 rafters need a 2x10 ridge). Order ridge stock long, 20 ft. at least, to make it easier to stack (build) the roof.

Sometimes when rafters are overspanned (too small to carry the roof load), they need to be supported in the center by a purlin. Again, check the plans for size, measure the lineal feet needed and order long lengths (such as 18 ft. or 20 ft.). Often builders will not order extra material for purlins, planning to use up any stock left over from joists or rafters. Codes may also require the use of 1x4 or 1x6 collar ties installed every 4 ft. on opposing rafters to help tie the roof structure together.

Trusses

When roof trusses are specified, check with the manufacturer to see how much lead time is needed to build and ship your order. Often this can be three or four weeks. Trusses can be ordered over the phone, but it is best to meet with a company representative so that no mistake is made in regard to number, style and size.

Fascia

The ends of rafter tails are often covered with a fascia board. This board should be long and straight, and of good-quality material because it is exposed. Frequently it is also 2 in. wider than the rafter, just like the ridge. Figure out the lineal feet required, add 5% extra and order long pieces. Some builders like to go to the elevation view on the plans, scale the length of all fascia boards and mark these lengths on the plan. This is a good idea especially when a building has barge rafters that are made from fascia board, which might be 22 ft. long, for example. If you order all 20-ft. stock, each barge rafter would have to be spliced. It is much easier to order longer stock, at least for the barge rafters. The plans will specify the type of material required, such as rough-sawn Douglas fir, redwood, cedar or pine.

Starter boards

When starter boards are used on rafters to cover open eaves, check the plans to determine the type and style of wood. Often it will be 1x6 pine shiplap with a V-joint. In humid areas, western red cedar is more common because it is more resistant to rot. If the overhang is 20 in., for example, it will take four rows of boards to cover it. Measure the lineal footage of the overhang, multiply this by four and add 10% extra. Exposed eaves can also be covered with finish-grade plywood. Again, with a 20-in. overhang, measure the lineal footage and divide by 8 ft., the length of a sheet of plywood, and by two because two 20-in. rippings can be cut from each 4-ft. wide sheet of plywood.

Roof sheathing

The amount of roof sheathing required is calculated from square footage. Multiply the rafter length by the building length and, unless starter boards will be used, include the eaves and any overhang that was created by barge rafters at the gable ends. Double this figure to cover both sides of the roof, and proceed from there just as you did for floor sheathing. Check the plans and/or code for thickness and grade.

Hardware

The most commonly used framing nails are 8-penny (8d) and 16-penny (16d) box nails with a vinyl coating (often called "sinkers"). The vinyl coating makes them drive easier and hold better, but it may not be a good idea to hold them in your mouth. It takes about 50 lb. of 16d nails and 12 lb. of 8d nails to frame 300 sq. ft. of house. So a typical 1,200-sq. ft. house can be framed with about four 50-lb. boxes of 16d and one box of 8d sinkers. You may have to supply all the washers and nuts to attach the sill to the anchor bolts.

The plans will note whether any framing anchors are needed, such as those used when a stronger connection is required between the rim joist and the sill, from post to beam or from rafter to plate, to hold things together in hurricanes or earthquakes. Metal angle braces are often used as permanent wall braces. Joist hangers at beams and headouts will also be needed, as will metal plate straps wherever top plates have been seriously weakened by cutting for plumbing or heating. In earthquake areas, codes often call for tiedowns, metal angles that bolt to the foundation and directly to a stud, although some people say that prayer works better. In high-wind areas, hurricane clips are often required. When beams or girders are built up from three or more pieces of lumber, they may need to be bolted together.

The completed list

Work carefully, check the plans, scratch your head, write it all down and submit your list with full confidence that you have overlooked something. The lumber list is, after all, an estimate of materials that will be needed for a particular job. No one expects an estimate to be perfect, just close. Remember that overestimating can be expensive. Lumberyards will take returns, but they may charge 15% to 20% to process the lumber back into their inventory. On the other hand, if you underestimate you can always order more.

It is the lumber company's responsibility to stack the material in the order you specify and to deliver it when requested. You don't want to deal with a company that puts sill stock at the bottom of a load when this is the first item you need to begin construction. It is always important to make sure that lumber loads are dropped close to where they will be used. Hauling lumber by hand from any distance consumes time that could be better spent framing the building. Many builders will have materials delivered as needed rather than all at once, especially if there is no way to make them secure at the job site.

The hardware required on a typical job includes nails, nuts, washers, joist hangers, plate straps and hurricane clips.

FRAMING FLOORS

Sills

Posts and Girders

Joists

Sheathing Floors

2

SILLS

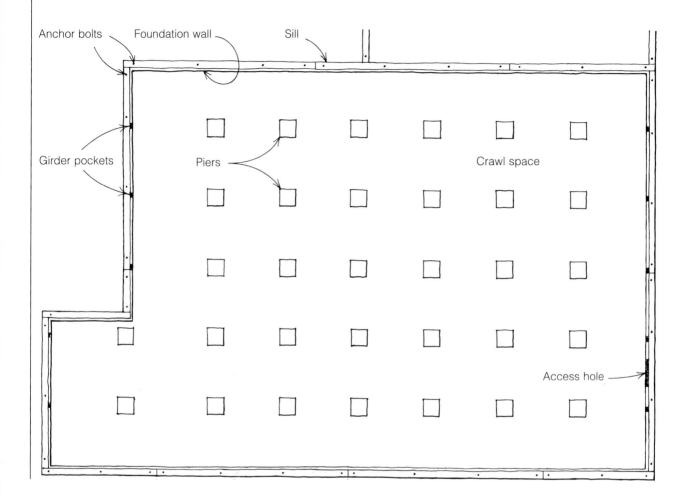

Anchor bolts   Foundation wall   Sill

Girder pockets   Piers   Crawl space

Access hole

Today many homes are built directly on a concrete slab, especially where winter temperatures do not fall much below freezing. Wooden floors take longer to build, but they are more resilient than slabs and they allow easy access to plumbing pipes, ductwork and electrical conduits. Whether you are building on a concrete slab, over a crawl space or over a full basement, the first wooden member that is laid down is called a sill or mudsill.

The sill is commonly a pressure-treated 2x4 or 2x6. It is usually attached directly to the foundation by anchor bolts that are embedded in concrete. Most codes require that anchor bolts be at least ½ in. by 10 in. and be located 1 ft. from each cor-

ner of the foundation, 1 ft. from the ends of each sill and a maximum of 6 ft. on center throughout. Regardless of its length, every piece of sill needs at least two bolts in it. The lumber used for sills often comes in 16-ft. lengths, which require a minimum of four bolts in each sill. These requirements can vary regionally, so check your plans or local building code if you are unsure. Anchor bolts are an important part of a structure; they help keep the house anchored in place during earthquakes, tornadoes and other natural and unnatural occurrences.

As the foundation is supported by the earth, so the house will be supported by the sill resting on and anchored to the foundation. If this first member is installed straight, square and level, everything that follows will go faster and better. Good workmanship generally costs less in the long run, and it doesn't have to take more time. There are many ways to do quality work quickly and efficiently, and that is what this book is all about.

Checking foundations

A good concrete contractor will leave the foundation square and level, but it never hurts to check before laying down the sill. If the foundation is a simple rectangle or square, it can be checked for square by measuring diagonally from corner to corner; if it measures the same both ways it is square.

If the foundation has offsets in it, for example, if it is shaped like an "L" or a "T," there is still an easy way to check all the corners for square. This is done by using the Pythagorean theorem: $a^2 + b^2 = c^2$. On the job this is usually called the 6-8-10 rule. Just measure 6 ft. from the corner along one side of the foundation, 8 ft. along the other side, and then check the length of the diagonal. If it is 10 ft., then the corner is square. It's that simple. It doesn't have to be perfect—you can do some correcting when you lay down the sills (see p. 34). You can use other multiples of 6-8-10, such as 3-4-5 or 12-16-20. On a big building with long walls, use as large a multiple as possible to avoid error.

Next check the foundation walls for parallel with a long tape by measuring across one end, then moving to the other end of the foundation and measuring again. If you come up with the same distance, the walls are parallel. Check this distance with the plans. Some error is permissible, depending on the size of the building. Walls that are out of parallel ½ in. in 10 ft. are worse than walls that are out of parallel ½ in. in 100 ft.

You need to check that the tops of the foundation walls are level. With a little practice, you can do this by simply sighting across the tops, but you can also use a builder's level with a laser or a water level (see the drawing on the next page).

Checking the Foundation for Square and Parallel

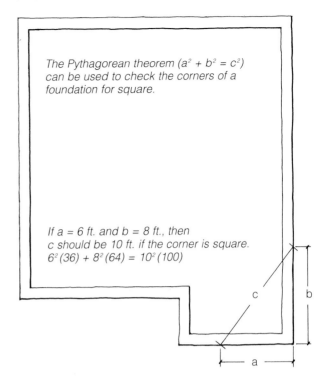

The Pythagorean theorem ($a^2 + b^2 = c^2$) can be used to check the corners of a foundation for square.

If a = 6 ft. and b = 8 ft., then c should be 10 ft. if the corner is square. $6^2 (36) + 8^2 (64) = 10^2 (100)$

The foundation is square if the two diagonal measurements are the same: A to B = C to D

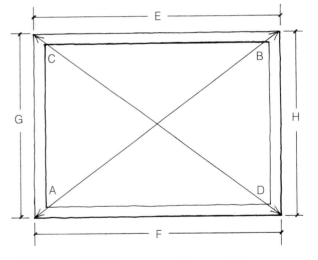

Walls are parallel if they measure the same distance across each end: E = F and G = H

Using a Water Level

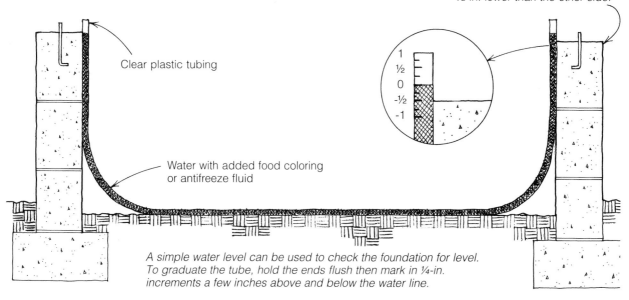

Clear plastic tubing

This side of the foundation is ½ in. lower than the other side.

1
½
0
-½
-1

Water with added food coloring or antifreeze fluid

A simple water level can be used to check the foundation for level. To graduate the tube, hold the ends flush then mark in ¼-in. increments a few inches above and below the water line.

Bent anchor bolts can be straightened easily with a piece of ¾-in. pipe.

Straightening bolts

Make sure that the anchor bolts are sitting straight up. If any of them have been bent or were put in crooked, you can straighten them with a length of ¾-in. pipe. Just slip the pipe over the bolt and bend it upright. Also, if the concrete contractor placed a bolt where one doesn't belong, break it off with the pipe by bending it back and forth a few times. Alternatively, you can use a reciprocating saw with a metal-cutting blade.

Scattering wood

An efficient framer doesn't carry, mark, cut, drill and attach one piece of wood at a time. Production framing means repetition: Do as much of one task as possible before going on to the next one. "Scattering" is one of those efficient procedures that production framers routinely employ. Before you do any measuring and cutting, carry and scatter all the sill stock, that is, place all the boards end-to-end around the foundation. Usually they can be laid roughly in position on the foundation near the bolts, which will make it easy to determine where they will need to be cut.

A 16-ft. long 2x6 can be heavy. Learn how to carry sill stock on your shoulder.

SAFETY TIPS

Handling pressure-treated lumber

Pressure-treated lumber contains some hazardous chemicals that ensure its durability. Manufacturers claim that if it has been properly treated and dried, pressure-treated lumber is relatively harmless to humans, but it is wise to take a few safety precautions when working with it.

• Wear gloves when carrying pressure-treated lumber.

• If you don't wear gloves, be sure to wash your hands before eating or drinking anything.

• Saw pressure-treated lumber outside, or wear a dust mask.

• Remove any slivers as soon as possible (tweezers are an important tool in a carpenter's toolbox).

• Don't burn scraps in your woodstove or fireplace. Combustion releases the toxic substances that are bonded in the wood.

Rest the sills on the foundation before cutting them to length.

Cutting sills

Many carpenters spend a lot of time measuring and marking lumber before cutting. This really isn't necessary. The sill stock is in position on the foundation so the building itself acts as a template, indicating where the cuts need to be made. With a little practice, you can cut them square, or at least square enough, by simply eyeballing, that is, visually aligning, the front edge of the saw table on your circular saw with the edge of the wood. Working this way squares up the blade with the wood and allows you to cut accurately across any board, even a 2x12, without using a square. Framing is not finish carpentry; it needs to be done accurately, but not perfectly.

A large amount of framing can be done by eyeballing. Carpenters need to train their eyes to do a lot of their measuring and marking for them and to learn to trust their judgments, which will improve with practice. Having to pull out a tape or square for every little measurement and cut is time-consuming and often unnecessary.

An efficient carpenter needs to learn how to cut lumber without always measuring and marking. When cutting sills, let the foundation edge be your guide for length. For a square cut, keep the edge of the saw table parallel with the edge of the lumber.

Positioning Sills

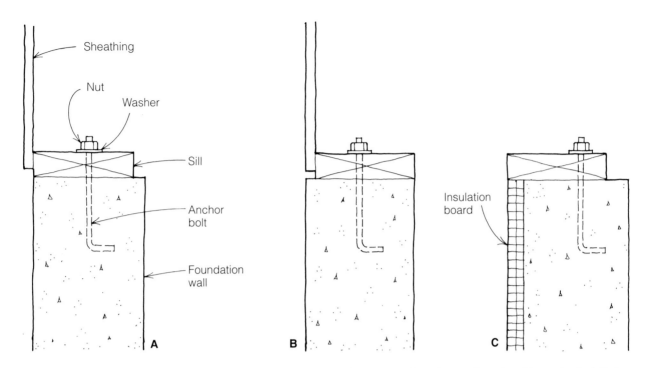

The sill may sit flush with the edge of the foundation (A), it may sit back from the edge to allow sheathing to be installed flush with the foundation (B), or it may overlap insulation boards applied to the exterior of the foundation (C).

When cutting the sills to fit the building, try to leave a bolt within 1 ft. of each end. If the concrete contractor has spaced the bolts properly, this shouldn't be difficult. If extra bolts are needed, an expansion bolt, also called a redhead or wedge anchor, can be inserted into the foundation after the concrete has hardened. These bolts are typically ½ in. by 6 in. and are fitted into a ½-in. hole drilled about 4 in. deep into the concrete with a heavy-duty drill or rotary hammer and masonry bit. Leave the nut on to protect the threads and drive the bolt into the hole. It will expand as it hits the bottom, securing itself in the foundation.

Marking bolt holes on sills

The next step is to mark where the holes will be drilled in the sill. This is a time to think ahead and check the plans to see what will be covering the exterior walls. The sill may be installed flush with the outside of the foundation, it may be set in from the edge of the foundation to allow room for sheathing or insulation board on the walls, or it may overhang the foundation to cover foundation insulation (see the drawing above). A plan detail should give you this information.

As long as the outside walls of the foundation are straight and parallel, marking bolt holes on the sill can be done by eyeball. If the sill sits flush with the outside wall, just set it on top of the bolts, sight down the outside of the foundation and hit it with a hammer right on top of every bolt. Striking the sill leaves a mark on the wood for drilling. If the sill sets back on the foundation wall or if the wall isn't straight, snap a chalkline on top of the foundation and align the sill with it. Doing this accurately takes a bit of practice but will save you time once you get the hang of it.

If you're building on a concrete slab, the bolt holes on the sill plate can be marked with a bolt-hole marker (see pp. 77-78).

To mark the sill for drilling, set it on top of the bolts, flush with the outside of the foundation, and hit the sill with your hammer over each bolt.

Even if the foundation walls are not totally straight and parallel, the sills can be. If the walls aren't parallel, snap chalklines that are parallel and use them as guides for placing the sills. Try to equalize any adjustments as much as possible. For example, if the foundation is 1 in. out of parallel, don't make all of the adjustment on one end. Instead, make ¼-in. adjustments on both sides at each end. Mark all sills before you start drilling.

Attaching sills

Now prop the sills up on the foundation wall or across a scrap of wood and drill the holes using a ⅝-in. or ¹¹⁄₁₆-in. bit with a ½-in. power drill. When all the holes are drilled, sweep any debris off the surface and place the sills over the bolts onto the foundation; you may need your hammer to persuade a few boards to fit. Put the washers on, then the nuts, and tighten with a crescent wrench, a socket wrench or, better yet, an impact wrench. (On a house being built on a slab foundation, the sill serves as the bottom plate of the framed walls. The nuts and washers are left off until after the wall is framed and raised.)

Fitting Sills to the Foundation

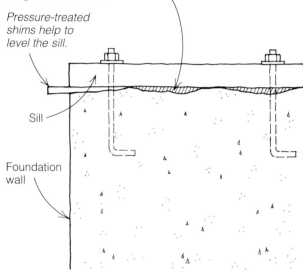

Fill gaps between the foundation and the sill with grout so that the sill will have full bearing to support the weight of the house.

Pressure-treated shims help to level the sill.

Sill

Foundation wall

The sill should fit tightly to the foundation wall and provide a square and level surface upon which to build a house. If the tops of the foundation walls aren't level, shims can be placed under the sills to bring them level. The resulting gap under the sill, and any dips in the foundation, must be grouted or "dry-packed," that is, filled tightly with a fairly dry mixture of concrete.

In many parts of the country, a thin layer of insulation is often laid between the sill and foundation. Some codes also require a sheet-metal termite shield to be placed between the two.

POSTS AND GIRDERS

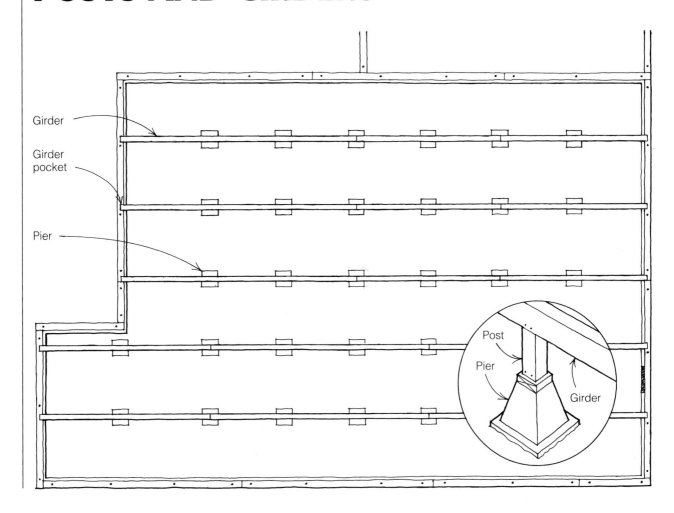

Girder

Girder pocket

Pier

Post

Pier

Girder

If you are working on a slab, you can start building walls right on the foundation. But when the house is being built over a crawl space or a basement, you have to build a floor system. The first step is to install posts and girders.

Girders are large horizontal beams that provide intermediate support for floor joists. Typically, they are supported on each end by girder pockets in the foundation. In between, they are supported by posts that are attached to pressure-treated 2x6 blocks attached to concrete piers, which can be purchased and set in wet concrete (pyramid piers) or formed and poured (square piers). Girders and joists may also be supported by underpinning walls or full load-bearing walls in a basement.

By breaking up the span of a building, girders allow the use of smaller joists. The shorter the distance between bearing points, the smaller the joists can be; conversely, the greater the span, the larger the joists need to be to carry the load. For example, if the span of a building or a room is about 20 ft., most codes allow the use of 2x12 joists at 16 in. on center spanning from wall to wall without any intermediate support. This method of joisting may be necessary over a basement or on second-story walls over larger rooms, or you can use wood I-beam joists (see p. 22). If you tried to span 20 ft. with 2x6 joists, they would sag in the middle. Two-by-sixes can be used as joists, however, when they are supported by properly spaced girders or load-bearing walls.

TYPICAL SPANS FOR FLOOR GIRDERS					
		Species: Douglas Fir-Larch Grade: No. 2 or better		Species: Hem-Fir Grade: No. 2 or better	
Size of girder	Spacing of girder	Partition walls above	No partition walls above	Partition walls above	No partition walls above
4x4	6 ft.	4 ft.	4 ft.	4 ft.	4 ft.
	8 ft.	3 ft.	3 ft.	3 ft.	3 ft.
4x6	6 ft.	6 ft.	7 ft.	5 ft.	6 ft.
	8 ft.	5 ft.	6 ft.	4 ft.	5 ft.
4x8	6 ft.	8 ft.	9 ft.	6 ft.	7 ft.
	8 ft.	6 ft.	7 ft.	5 ft.	6 ft.

Girders over a crawl space usually span the length of the building and are often 4x6s or two 2x6s nailed together. Four-by-six girders spaced 6 ft. on center typically need to be supported by a post every 6 ft., but this can vary with the type of wood used and the load that the system will bear. The span is often greater over a basement than over a crawl space, in which case the girders or the joists will have to be larger. Check the plans for lumber size and type.

Post length

In areas where termites or moisture are serious problems, the posts in a crawl space may be made from pressure-treated wood. Codes generally require that girders in crawl spaces be at least 12 in. off the ground, and the posts are usually 1 ft. to 2 ft. long. Posts in a basement are much longer.

The exact length of each post needs to be determined. First string a dryline tightly from sill to sill over the tops of a line of piers. Then place a scrap piece of girder stock on the pier (check the plans for girder size). The distance between the string and the

Measuring Post Lengths

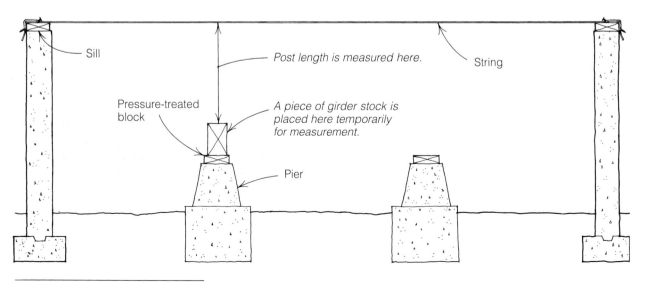

Sill

Post length is measured here.

String

Pressure-treated block

A piece of girder stock is placed here temporarily for measurement.

Pier

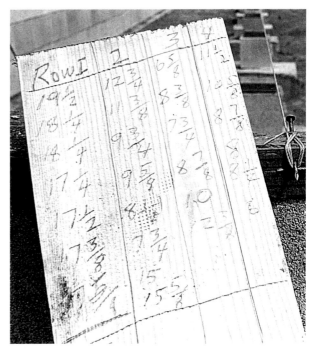

Post lengths should be recorded on the pier blocks and on a cutting list. Keeping a cutting list allows you to cut all the posts at one time.

top of the girder stock is the length of the post for that pier. Repeat this process for each pier. Write the length of the post on top of the pier block and also keep a cutting list of the lengths needed on a scrap of wood.

Some of the piers may not be perfectly level. Check them with a level or your eye, if you trust it. A post placed on a pier that isn't level prevents the girder from resting with full bearing on each pier. When you run into this situation, cut the post so as to compensate for the angle of the pier. A quick way to do this is to measure the long distance and the short distance from the dryline, transfer both measurements to the opposing sides of the post, and cut the post at the proper angle. An experienced carpenter can measure to the short point and then make the cut by eye.

Posts are usually cut from 4x4 stock, but check your plans. They can be cut to length with a circular saw, a chopsaw or a radial-arm saw. Gather a supply of stock and your cutting list and cut all the posts at one time. Write the length on each post and scatter them to their appropriate piers.

4x4 posts can be cut accurately with a circular saw by keeping the saw table parallel with the edge of the post. Make the first cut and, keeping the saw square, finish by turning the post.

Girder Placement

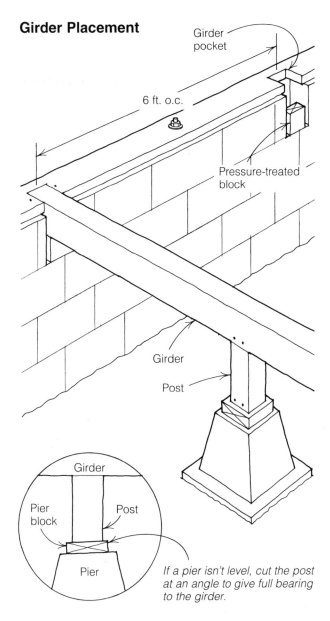

Girder pocket

6 ft. o.c.

Pressure-treated block

Girder

Post

Girder

Pier block

Post

Pier

If a pier isn't level, cut the post at an angle to give full bearing to the girder.

Each post is nailed to the pier block with three 16d or four 8d nails.

Framers often leave the dryline used to measure post length in place until the posts are nailed to the piers. It serves as a guide to ensure that posts are nailed to the piers in perfect alignment, offering full bearing to the girder. Grab your hammer (the sidebar on the facing page shows you the correct way) and toenail three 16d (two on one side, one on the opposite side) or four 8d nails through the posts into the pier blocks.

Girders

With the posts all nailed in position, it is time to scatter the girder stock. Since piers are often 6 ft. apart and girders need to break over posts, girder stock is often 12 ft. and 18 ft. long. Use straight stock for girders so that the floor joists will have a good level surface to rest on.

For standard 4x6 girders, the pockets in the concrete foundation are typically 4½ in. wide, 5½ in. deep and about 4 in. long. (On a block foundation, the pockets are normally the depth of the block.) Girders rest on a piece of pressure-treated 2x stock

Proper hammering

Every carpenter should learn how to swing a hammer safely. Over time, an improper grip can lead to tendinitis and sore wrists. Grab the handle securely with the entire hand, wrapping the thumb around it as shown in the photo below. This way, the two middle fingers act as a pivot as the hammer is moved up and down in the process of driving nails. A good nail driver learns that even though the entire arm is involved in hammering, it's the snap of the wrist that drives the nail home.

Eric Haun

Holding a hammer with a full grip will help prevent tendinitis and make it easier to drive nails.

in each pocket to keep untreated wood off the concrete. A notch has to be cut into the sill to allow the girder to slip down into the pocket, leaving it flush with the top of the sill. If the pocket is too deep, place thicker pieces of wood in it for the girder to rest on. If the pocket isn't deep enough, use a thinner piece of wood or trim a bit from the bottom of the girder.

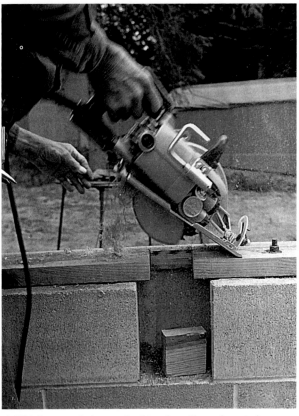

A 3½-in. notch is marked in the sill at the girder pocket. Make two side cuts and a plunge cut to remove this cutout and make room for the girder.

The pocket needs to be wide enough so that the girder has ½-in. clearance between it and the concrete on both sides. In some regions, building codes also require a metal termite shield to be wrapped around the end of the girder.

Girders must break over posts. If the lumber was ordered carefully and the piers were laid out accurately, the girders should fit exactly. If the girders have to be trimmed, cut them so that they break in the middle of the post, providing equal bearing to each piece of girder. Secure the girders with three 16d or four 8d toenails into the top of the post. Check the plans to see whether girder breaks need to be spliced together with a metal strap or a plywood gusset (plate).

If the posts in a crawl space are more than 3 ft. tall, most codes require them to be braced with pieces of 1x4. Run these braces from the bottom of the post at a 45° angle up to the girder. The code may require a brace running both ways. Nail the braces with five 8d nails on each end. In a basement, tall bearing posts are often part of an interior wall that divides the space into rooms. These walls tie into the foundation and the girder-joist structure. They are often sheathed on one or two sides with plywood and add structural strength to the building.

Girders are nailed securely to each post with three 16d nails or four 8d nails, helping to tie the entire frame to the foundation.

Bracing and Splicing Girders

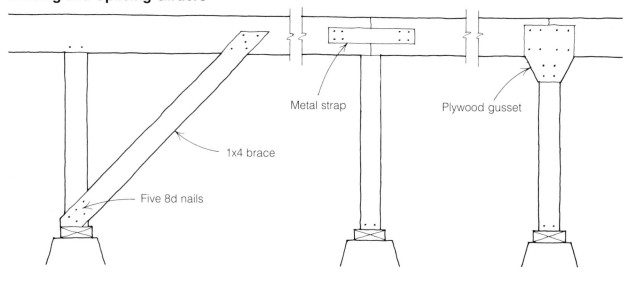

1x4 brace

Five 8d nails

Metal strap

Plywood gusset

JOISTS

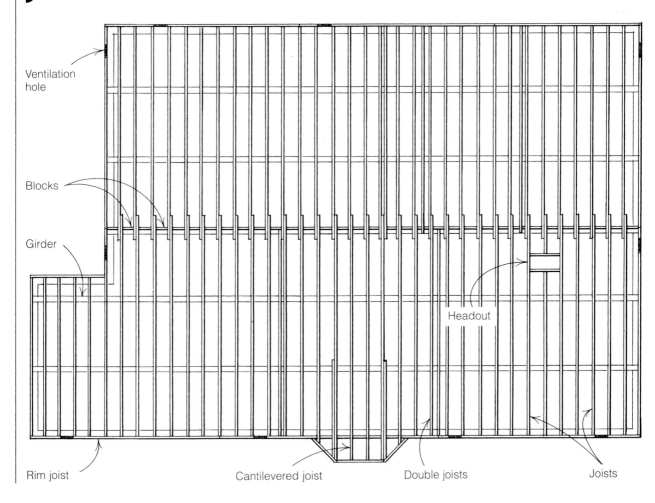

Ventilation hole

Blocks

Girder

Rim joist

Cantilevered joist

Headout

Double joists

Joists

First-floor joists are placed on edge across the sills to provide support and a nailing surface for the subfloor and a platform for walls. On a small building, single joists may be able to span the entire foundation. On larger buildings, the joists will span from the exterior wall to a girder, beam or bearing wall, where they will either butt or lap over the top. At the sills, they are supported and held upright by rim joists. Joist size can vary from 2x6s over girders in a crawl space to 2x10s, 2x12s or prefabricated plywood I-joists over basements and second floors.

Large beams can be used to allow larger spans, and thus larger rooms. The beam, supported by posts, is usually installed at the same height as the joists. The joists are then hung from the beam, sup-

ported by metal hangers. Check the plans for specific information on beam and joist size, length, spacing and the direction the joists will run.

One problem with beams that are installed flush with the joists is that much of today's framing lumber has a fairly high moisture content, and joists, being smaller than beams, will dry out and shrink faster. When plywood sheathing is nailed to a newly installed joist-beam floor system everything will be level and straight initially, but over time the joists may dry and pull away from the sheathing, creating the all too common squeaky floor. Eventually, but it may be years, the beam will also shrink and the floor will be level again. One solution is to install plywood I-joists, such as TJI joists (see p. 22), or an-

Plywood I-Joist

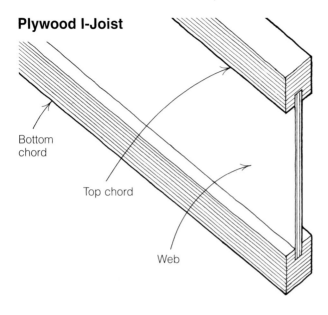

Bottom chord

Top chord

Web

other form of engineered wood product. These joists will shrink very little. Beams can also be run low, like typical girders, with the joists lapping over them. This system takes up more vertical space, but that can be an advantage if the intention is to leave an exposed beam for aesthetic purposes.

Joist systems are often taken for granted: out of sight, out of mind. Yet between the joists are pipes, ducts and wires to carry water, heat and electricity to all parts of the house. Filled with insulation, joist systems help keep homes warm in the winter and cool in the summer. Floors have to be built strong enough to support people as well as refrigerators and grand pianos. Codes often require that they be able to support 40 lb. per sq. ft.

Rim joists

The rim joist is a continuous piece of lumber, the same size as the floor joists, that is nailed into the sill or wall plate around the perimeter of the building. Floor joists are nailed to the rim to help hold them upright. Traditionally, the rim supporting common joists was formed by blocks. Floor joists were cut full length and nailed on edge to the foundation sill or wall plate, with blocks placed between them to provide stability and proper spacing. This is a good construction technique, but cutting and nailing individual blocks is much more time-consuming than using a continuous rim.

If you are working over a crawl space close to the ground, the most efficient method for joisting is to nail the rim on first. Scatter long, straight rim stock around the building. Then, holding the rims on edge flush with the outside of the sill, nail them down with 16d toenails every 16 in. on center. Some codes require that the rim be secured to the sill with additional small framing anchors nailed to both.

Joisting for a second floor over living-space walls is a little different than over a crawl space. These walls have to be plumbed and lined (see pp. 116-124) before joists can be scattered. Once the walls are ready, carpenters like to scatter all the joists flat across them first. When the joists are positioned, they can be pulled back from the wall edge at least 1½ in. to leave room to nail on the rim joist. This way, the flat joists provide a safe surface from which to work.

Layout

Once the rim is nailed on, mark the joist layout. Joists are most often placed 16 in. o.c., but check the plans to make sure. They can be spaced 24 in. or even 32 in. o.c. if heavier floor sheathing is used. Plans also indicate the direction joists run. Some may run parallel with the outside wall, some perpendicular. When joisting over a basement or for a second floor, take the floor-framing plan, walk through the house and mark with keel the direction, length and number of joists needed on the floor below. Note also the size, length and location of any beam that is to be inserted to break up the span. This way you won't have to be constantly referring to the plans as you carry and position all the joists and beams on the walls over the rooms below. Any beams that are needed to break up the spans should be installed at this time before continuing with the joist layout.

Most construction manuals indicate that the first joist should be positioned 15¼ in. from the outside edge. This is so that when the floor is sheathed with 4x8 plywood, the plywood panel will have full bearing on the rim joist and split the center of the joist 8 ft. away. But this step can easily be eliminated. If you lay out the first joist at 16 in. o.c. rather than 15¼ in., the end of the plywood will split the center of the rim joist just like it does for every other joist throughout the building. In other words, it will have ¾ in. bearing on the rim joist instead of 1½ in. No structural change is made and, once it is covered by the bottom plate of a framed outside wall, it won't even be visible. Simplifying the layout saves time, and that's what efficient carpentry is all about.

Hook a long measuring tape on the end of the rim and make a mark on top of the rim every 16 in. down its entire length. Use a blue or red piece of keel, which leaves a much more visible mark than a pencil. After making each mark, put an X or a simple dash alongside of it to show which side of the line the joist will go. (The X is somewhat traditional, but a dash is quicker.) A good carpentry tape will have a mark every 16 in. to facilitate layout.

If one joist spans from rim joist to rim joist, the layout will be identical on each rim. But if joists lap over a girder or wall, the opposing rims need to be laid out a little differently. The on-center marks on each rim will be the same, but the joists will go on opposite sides of the line. That way, the joists will lap each other over a girder or wall, where they will be stabilized with 2x blocks. Because these blocks will position the joists, you don't need to run a layout on the girder or wall. Beams that have been installed need to be laid out and the joist hangers nailed in place.

Working from the plans, be sure that you lay out the location of any stairwells, fireplace or chimney openings, or large access holes so that the main supporting joists can be nailed in the proper location. Locate the openings on the actual structure by transferring the measurements on the plan to the sill or rim joist. If the plans call for a 36-in. set of stairs, for example, leave 37 in. for the rough stairwell, allowing ½ in. for drywall on each side. Smaller holes for plumbing or heat vents can be framed in after the joists are nailed in place. If the plans call for a cantilever (that is, joists projecting beyond the foundation or wall line) for a bay window, room extension or balcony, the exact location and width need to be marked on the sills or rim joists. The rim will need to be cut to allow longer joists to project out over the edge of the building.

Joist Layout

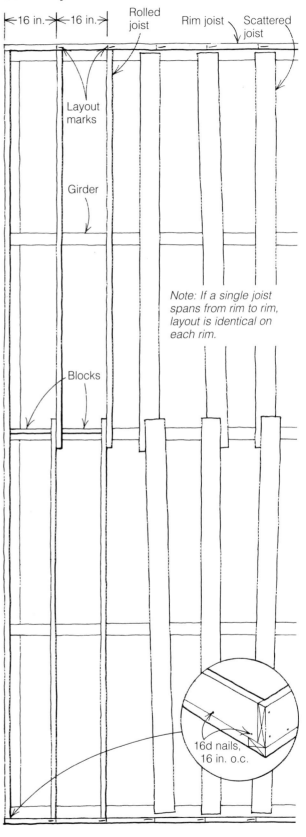

← 16 in. → ← 16 in. →

Rolled joist

Rim joist

Scattered joist

Layout marks

Girder

Blocks

Note: If a single joist spans from rim to rim, layout is identical on each rim.

16d nails, 16 in. o.c.

Carrying lumber

Frame carpentry requires carrying a lot of lumber. Because of its size and moisture content, much of it is heavy. Experienced carpenters have been known to identify only three types of lumber—redwood, deadwood and leadwood. Wet 2x12s fall into the latter category. Learn to lift with your legs and arms and not with your back.

People just starting in the trade most often carry lumber in their hands held down at their waist. If you have to do this for any length of time it is likely to cause undue strain on the lower back. Boards are most easily carried on the shoulder. It's a trick that's easily learned: Grab a joist at the balance point, and in one fluid motion lift it and flip it so that it lands gently on your shoulder. The weight of the joist is now directly on the body frame.

Joist stock can be heavy. Experienced framers learn how to lift joists with their legs and carry them on their shoulders.

Scattering joists

Having to carry all the joist stock around the foundation or second floor will make you, and your back, really appreciate a well-placed lumber drop (see p. 26) or a forklift. Carrying joists a long distance can wear you out fast. Check the plans to determine what lengths of joist are needed for each span. Longer ones will be needed for cantilevers that hang out over the building line. Don't worry about joists that are a little long; they will be cut to size later. Keep an eye open for joists that are badly bowed or twisted or have large knots in them. These can be set aside to be cut up and used for blocks or short joists later on.

If you are scattering joists for a second floor, don't rest too many up against a wall—their weight can push it out of plumb or alignment. Instead, put one end of the joist up on a wall, then, standing on a short stool, a work bench or a bucket, place the opposite end on the parallel wall.

Cutting and rolling joists

With the joists scattered and any necessary rims and beams in place, it's time to cut and "roll" the joists. (Rolling means nothing more than setting them on edge and nailing them on.) As a general rule, lapped joists should lap at least 4 in. If they lap more than 12 in., cut off the excess and use it for blocking. Cutting joists to length is easy to do, because once they are in place, the building itself does the measuring. Just eyeball and cut.

Joists that butt beams are rolled on edge with the crowns up and dropped into the hangers. Secure them first with a 16d toenail driven through the joist into the beam. Along with the hanger nails, this toenail should prevent the joists from moving in the hanger and causing floor squeaks later on.

When joisting a second floor, joists that extend from a beam to an outside wall are often cut to length after they have been rolled in place. If the cut is flush with the outside, sight down the joist and begin the cut straight up from the wall. Cut until the saw table hits the top plate, reach down, pick up the joist a bit and finish the cut. If the cut is 1½ in. back from the edge to leave room for a rim joist, sight down and hold the left side of the saw table (on a worm-drive saw) back about ½ in. from the inside of the wall and make the cut. On a 2x4 wall, this will leave the joist bearing 2 in. on the wall with 1½ in. for the rim joist. It takes a little practice to make this cut accurately, but once you have it down you will be surprised at how much it will speed up joisting. Just remember, when working on a second floor, watch for fellow workers below.

An easy way to roll the joists is to stand on the flat joists, reach down and grab one and sight down it to check the crown, that is, the bow along the edge of a joist. Crowns should be placed pointing up because when the joists are sheathed and begin bearing a load they tend to straighten out; or at least they won't sag. When joisting over girders that are close together, this practice doesn't matter so much because the joists will be held straight by the girders once they are nailed in position.

Floor joists should be installed with the crown up.

Once all the joists have been scattered, they can be secured to the rim joist with 16d nails.

To secure joists to a rim, line them up with the layout and drive two 16d nails through the rim, directly into the end of each joist. Nails can be driven with a hammer or with a nailer. Drive one nail high and one low to help keep the floor joists upright and stable. Learn to set up a rhythm when you work, a sort of dance. With practice, your body will begin moving comfortably from joist to joist. Keep your eye open for any special layout, such as for a cantilever or for a stairwell. There's no reason to nail in joists where they aren't needed. Before long you should be able to look back and see a "sea of joists." It feels good to see all you've accomplished.

Cutting and nailing blocks

In an earthquake, unblocked joists can roll over flat. Blocking helps prevent any joist rotation and adds strength and stability to the entire floor. Blocking is usually required between lapped joists over a girder or wall, and some codes require all joists (lapped or not) to be blocked over all bearing points. If the blocks are cut accurately, they will automatically space the joists correctly. On the standard floor with joists spaced 16 in. o.c., the blocks that go between single joists are 14½ in. long. Blocks between lapped joists are 13 in. long (see the drawing on p. 43).

Before you cut any blocks, however, check to see how thick your joist stock is running. You might find that it's a bit thicker or thinner than 1½ in., in which case you would need to adjust the size of your blocks accordingly.

Blocks can be marked to length quickly using a framing square.

The preferred tool for cutting blocks is the radial-arm saw. Once it is set up, a lot of blocks can be cut in a short time. But you can also cut blocks quickly with a circular saw. This is a good time to use up some 2x scraps. If you're cutting blocks with a circular saw, you can lay them out quickly with a framing square. Assuming that you are cutting 14½-in. blocks, align the end of the 2x with the 14½-in. mark on the blade of the square. Then draw a line across the 2x using the inside of the tongue of the square as your guide. Using this method, you can mark hundreds of blocks in very little time. When it's time to cut them, hold the sawblade to the right of the line ("leave the line") to ensure that each block will be the same length.

The blocks can now be scattered near to where they will be nailed on. You can lay pieces of 1x6 across the tops of the joists to give you a surface on which to set the blocks.

As a standard practice, when nailing blocks between lapped joists, nail them flush with one side or the other of the girder or wall. This will make it easier to add extra joists later that may be needed to support walls above (for more on this, see p. 51).

Be careful to start correctly. You want to make sure that the joists maintain the same on-center layout at the blocks as they do at the rim joist. After nailing in a few blocks, use your measuring tape to check for accuracy. Begin nailing in blocks by setting the first one on edge, flush with one side of the girder or wall. Drive a 16d toenail through the top of the block into the joist. Draw the next joist up to the block and drive two more 16d nails through the joist into the block. Then pull the lapped joist up against the first joist and nail these two together, again with two 16ds. Secure this joist to the girder or wall with another 16d toenail angling down through the joist. Then grab another block and repeat the process. Once you reach the end, turn around and drive a 16d toenail through the back side of every joist into the girder or wall. Always nail all the joists and blocks in one direction before turning around. Every joist should be toenailed to its support on both sides. Never put a nail in the top of a joist because this could cause you to dull a sawblade when cutting floor sheathing.

Finally, each joist needs to be secured to the sill or wall plate at the rim. Do this by walking one way, driving a 16d toenail through each joist into the sill or plate. When you reach the end, turn around and repeat the process on the other side of each joist. Do the same over every girder or interior wall. Before nailing to these intermediate supports, check that the joists are running straight from the rim to the lap blocks.

Special length blocks ("specials") may be needed in some locations, for example, between the rim joist and the last lapped joist. Usually these can be cut by laying a piece of joist stock flat across the space it will fill, butting the end against the adjacent joist, sighting down and cutting.

Nailing Blocking at Lapped Joists

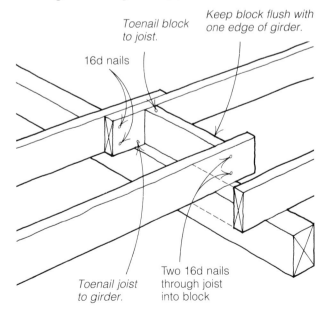

Toenail block to joist.

Keep block flush with one edge of girder.

16d nails

Toenail joist to girder.

Two 16d nails through joist into block

Nail all the blocks and joists in one direction, then turn around and nail from the other side.

Headouts

Frequently holes need to be cut in the floor frame to allow for such things as stairs, chimneys, plumbing runs, heating ducts and attic access. These openings are made by cutting ("heading out") joists. When any joist is cut, however, there is a loss of structural strength around the opening. The opening must be framed back in so that this lost strength will be restored. In general, building codes state that when the header joist exceeds 4 ft., both the side (trimmer) joists and the end (header) joists have to be doubled. Many framers double the trimmer and header joists any time they have to cut more than one joist. Double joists are nailed together every 16 in. to 24 in. o.c. depending on local requirements. When joisting with wood I-beam joists, all headouts have to be planned and prepared for in advance. Any on-the-job cutting or notching can seriously weaken these prefabricated joists.

The location of the stairwell was marked from the plans during layout. The length of the opening should also be on the plans. A typical straight flight of stairs running from first floor to second, with 8-ft. high walls, requires a stairwell opening between 120 in. and 130 in. long.

The precise location of many of the smaller headouts is not always indicated on the plans. For example, the tub trap and toilet-closet bend, which will be installed by the plumber, have to be located exactly, but you're not likely to find the necessary dimensions on the plans.

It's easy to locate a bend for a toilet in the corner of a bathroom. Measure 15 in. off the inside of the side wall and 12 in. off the back wall and center an opening at least 14-in. square in that location. The trap for a corner tub is located by measuring 15 in. off the inside of the side wall and 6 in. off the back wall, and centering a 14-in. square opening in that location. If a joist is in the way, make marks on top of it, leaving an extra 1½ in. on each end for the header joist, so that you will know where to make the cuts. These holes do not have to be exactly 14 in. square; it's actually better, when possible, to make them 3 in. or 4 in. larger. Dimensions can vary a little regionally, so check with your local plumber, heat and air-conditioning installer and other trades-

Framing Headouts

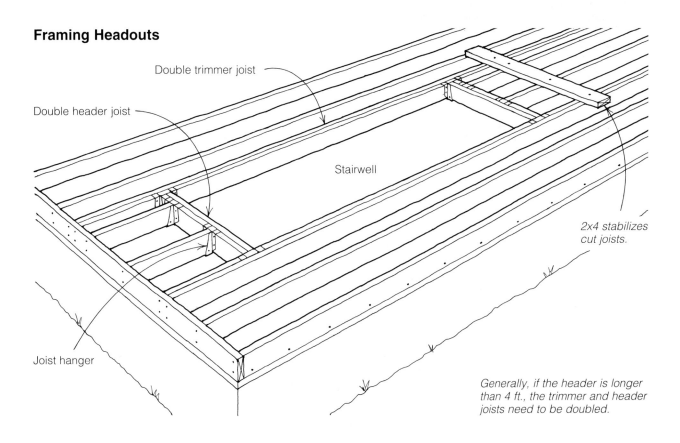

Double trimmer joist

Double header joist

Stairwell

2x4 stabilizes cut joists.

Joist hanger

Generally, if the header is longer than 4 ft., the trimmer and header joists need to be doubled.

This opening, or 'headout,' in the floor has been framed in to allow for plumbing for a toilet.

people for the exact location and size of headouts they will need to run their pipes and ducts. Anytime you have to head out for a heater vent, make sure that no wood will be within 1 in. of the pipe to minimize the risk of fire.

Before cutting any joists, support them temporarily by nailing a long 2x across their tops. Standing on and cutting off unsupported joists spells trouble —the fall may not hurt you, but the landing will. After cutting the joists, attach the supporting header with two 16d nails plus a joist hanger at each intersecting point. Framers prefer the "R" type hanger, which hooks over the supporting joist, because it takes fewer nails to attach it. The "U" type hanger does the same job but takes more nails.

Joist hangers should be nailed in place with short, hardened hanger nails, unless the plans specify that you can use regular 8d nails. Many framers don't like to nail off hangers because it's hard to get hands, nails and framing hammer between joists at the same time without banging a finger. Some framers use an air-operated nailer (available from Danair, Inc., Box 3898, Visalia, CA 93278) to drive the short hanger nails, and these nailers, which fit in your palm, work quite well, especially if you have thousands of nails to drive.

The joist hanger on the right hooks over the supporting beam or joist and requires fewer nails to install than the hanger on the left.

Double joists

Double joists are also needed beneath "parallel walls" (overhead walls that run parallel to the joists) to carry the added weight. Some codes require a double only under load-bearing walls. Walls running at a right angle to the joists need no additional support. Traditionally, carpenters laid out and installed these doubles while nailing in all the regular joists. This is a legitimate way to joist, but it is more efficient to joist straight through on regular layout and then go back and install the double joists.

The location of parallel walls can be found on the floor plan. Transfer all parallel-wall dimensions to the rim joist and girders to locate the area that needs an extra joist. The double joist needs to be moved 2 in. or 3 in. to the right or to the left of this mark so that it won't fall directly under the wall but will leave room for conduit or pipes to be run from below up into the wall.

If the lap blocks were kept flush with one side of the girder or supporting wall, adding the double joists is easy work. Depending on which side of the lap block the double joist goes, you either cut the joist to rest on the girder or wall, which provides sufficient bearing, or you cut out a 1½-in. section of the block so the joist can slip down into this slot. Had the lap block been nailed in the middle of the girder there would have been insufficient bearing on either side of it, and you can't cut out a slot in a centered block with a regular circular saw. If you run the doubles this way you won't have to come back and cut special-length blocks at these locations.

Blocks often have to be cut to allow for a double joist. This is much easier to do if the block is installed flush with the edge of the supporting girder or wall.

Doubling Joists under Interior Walls

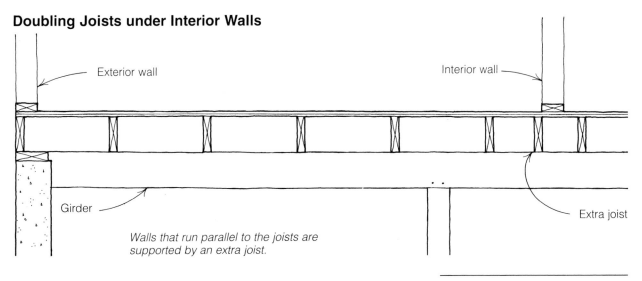

Exterior wall

Interior wall

Girder

Extra joist

Walls that run parallel to the joists are supported by an extra joist.

Cantilevered joists

Long joists are frequently used to cantilever beyond the foundation to provide support for a deck or balcony, room extension, bay window or even another story. The location of any cantilevers was marked during layout. The plans will indicate how far the joists cantilever beyond the building line and, if the overhang is long, whether or not the joists need to be doubled to support the weight they will carry.

If the cantilevered joists run perpendicular to the regular floor joists, they usually are attached with joist hangers to a double joist in the floor frame. Codes normally stipulate that the cantilever run at least twice as far into the building as it overhangs. Thus, a 12-ft. joist could be cantilevered 4 ft. But these rules depend on the overall length of the cantilever, the load it will carry and the size of the joists. Follow the directions of the plans or local building code.

To work efficiently, use somewhat longer than necessary joists on a cantilever and then cut them to length all at one time after they are installed. Check the proper length on the plans and snap a chalkline across the joists. Be sure to cut off an extra 1½ in. to allow for the addition of a rim joist. Cut them to length by walking on the joists and cutting square down the joist with a circular saw. By now you should have learned how to cut simply by eyeballing, thus saving the time needed to measure and mark each board separately. If the joists are not strong enough to support your weight while cutting, lay another joist across them to use as a catwalk. Finally, nail the rim joist to the ends of the cut-off joists.

Cantilevered joists for balconies and decks can be ripped to a slight slope to ensure that water will run off away from the house. You can make a template to speed marking the joists, as shown in the drawing below. The deck or balcony should be at least 1 in.

These joists are cantilevered for a bay window.

Cantilevered Joists

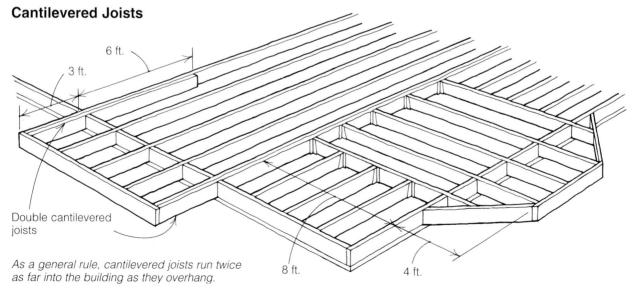

6 ft.

3 ft.

Double cantilevered joists

8 ft.

4 ft.

As a general rule, cantilevered joists run twice as far into the building as they overhang.

lower than the inside floor, and it should slope at least ¼ in. to the foot. For example, if a balcony is to be 4 ft. wide, the template should drop 1 in. at the wall and then taper another 1 in. to the end, requiring a rim joist 2 in. smaller than the actual joist.

It is easiest to mark and rip the joists before they've been rolled and nailed on. Let the building be your guide for length. Mark all the joists with the template and then rip them. Once the joists have been rolled into place and blocked at the building line, they can be cut to length to form the deck. Nail the rim on and it's ready for sheathing.

Crawl-space ventilation

When building over a crawl space, vents are often installed in the foundation to allow moisture to escape and air to circulate. If there are no vents in the foundation, they can be cut into the rim joists. The standard screened vent will fit between two 2x6 joists, so all you need to do is cut the rim joist out at this point. The cut-out block can be nailed in flat between the joists to help support the floor sheathing. If the joists are larger than 2x6, cut a 5½-in. by 14½-in. hole low on the rim joist, leaving wood across the top to carry the sheathing.

Tapering Joists for Exterior Balcony or Deck

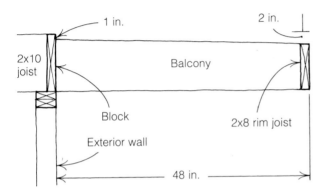

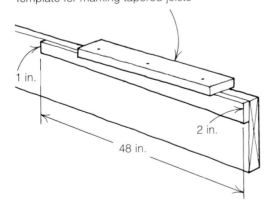

Template for marking tapered joists

Crawl-Space Ventilation

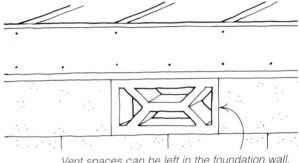

Vent spaces can be left in the foundation wall. If not, they need to be cut out of the rim joist.

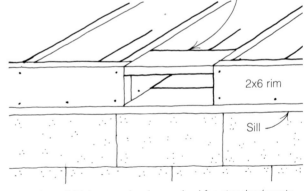

A 5½-in. x 14½-in. opening is required for standard vents.

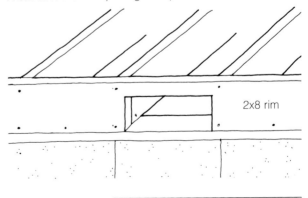

To ensure that the crawl space has adequate cross ventilation, many codes call for 1 sq. ft. of vent for every 150 sq. ft. of under-floor area. As a general rule, cutting a standard vent 2 ft. from each corner and one every 6 ft. on at least three sides of the building will meet code requirements. Structurally, it is not a good idea to put a vent underneath an exterior exit, and for aesthetic reasons most builders try to leave vents out of the front of the building.

Bridging

In residential construction, bridging (blocking joists at midspan) has pretty much become a thing of the past. It has been found that bridging provides little extra stability to a floor that is sheathed with plywood, and codes in many areas no longer require it on joists 2x12 or smaller. Bridging can consist of solid blocking or wood or metal cross bridging. Solid blocking, cut from joist material, is the fastest to install. When bridging is called for, it is spaced every 8 ft. or less midspan between the walls. When using solid blocking, stagger blocks so that both ends can be easily nailed with two 16ds driven through the joist into the block; this is much faster than toenailing. Drive the top nail about 1 in. down and the bottom nail as low as possible.

Drywall backing on top of walls

When joisting for a second floor or any ceiling that will be covered with drywall, backing has to be nailed on all walls that run parallel to the joists. On an outside wall this is done by nailing a 2x4 flat on top of the wall and letting it hang over into the room below (drawing A at right).

Interior walls may need a double joist if another wall is directly above, which will also serve as drywall backing below (drawing B). Otherwise, a 2x6 (drawing C) or two 2x4s side-by-side (drawing D) can be nailed flat on the double top plate for backing on both sides of the wall.

Nail the backing down at least every 16 in. with 16d nails. This is a good place to use up shorter pieces of 2x stock, crooked studs, badly crowned joists and lumber with large knots. If a parallel wall is long, keep it straight and plumb by nailing a 2x block, 14½ in. long, over the backing and between the adjacent joists.

Drywall Backing

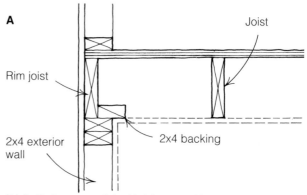

A

Rim joist

2x4 exterior wall

Joist

2x4 backing

Walls that run parallel with joists need backing for ceiling drywall.

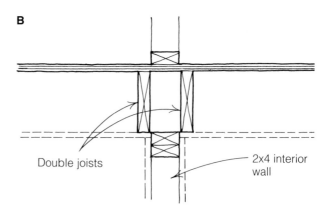

B

Double joists

2x4 interior wall

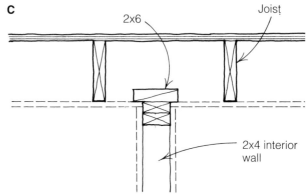

C

2x6

Joist

2x4 interior wall

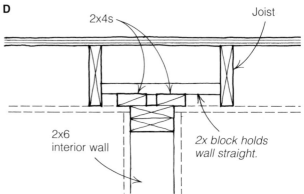

D

2x4s

Joist

2x6 interior wall

2x block holds wall straight.

SHEATHING FLOORS

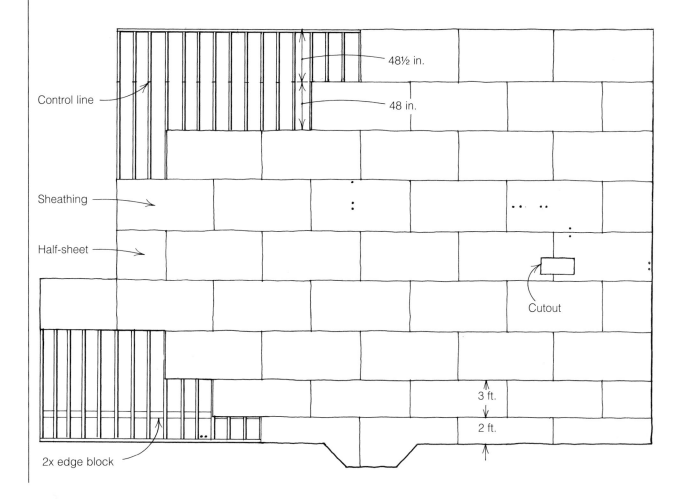

Control line

Sheathing

Half-sheet

2x edge block

48½ in.

48 in.

Cutout

3 ft.

2 ft.

Once all the joists are in position, a subfloor (also called a rough floor) can be laid. Most builders like to have the rough plumbing and ductwork installed before sheathing, but this can be done later. The subfloor is a sheathing that covers the entire floor, helping to tie all the joists together and providing a platform upon which to lay out and build walls. If you are building over a crawl space, it's a good idea to clean any debris, especially wood scraps, off the ground before laying the sheathing. Wood left under a building can attract termites to the area.

Sheathing materials

For many years the most common subfloor material was 1x6 lumber, which was often laid diagonally across the joists. There was a lot of waste with this application because each board had to be cut diagonally. Also, if the floors were going to be finished with carpet or vinyl, they had to be sheathed again with a plywood or particleboard overlay. Today, the subfloor is generally composed of a single layer of plywood or oriented strand board (OSB). Common sheathing materials are 4-ft. by 8-ft. sheets of ⅝-in. or ¾-in. Douglas-fir tongue-and-groove (T&G) plywood or OSB, made with an exterior-grade glue. Straight-edged sheets can be used, but the tongue and groove unites the sheets at the edges and decreases deflection between the joists.

OSB is made of directionally oriented strands of wood that are layered perpendicular to each other, much like plywood. These layers are then bonded, hot pressed and cut into panels. OSB is structurally strong and works very well for floor sheathing since it is available with a tongue and groove. But OSB can be susceptible to swelling when exposed to high humidity or long-term moisture, especially at the edges. Check to see if it is being used successfully for subflooring in your area.

Laying the first row of sheathing to a control line ensures that it will be straight.

Layout

Sheathing is laid with the long edge perpendicular to the joists. Begin on a side of the house that has a long straight run. It is best to work off a control line rather than the edge of the floor. On each end of the building, measure in 4 ft. ½ in. and snap a chalkline across the joists. The extra ½ in. allows for the tongue and any slight variation in rim-joist alignment. By following this line, the first row of sheathing will be absolutely straight, which will make the rest of the job much easier.

Edge blocking

If you are using straight-edged sheathing, the plans may call for the use of edge blocking, a row of blocks between joists to provide more structural stability to the floor. Edge blocking is generally not needed if T&G sheathing is used. When blocks are required, snap a control line 4 ft. ½ in. in from the edge of the building and then a line every 4 ft. across the building. Next cut a stack of 14½-in. (assuming a 16-in. o.c. layout) 2x blocks (this is a good time to use up some scrap) and lay them on 1x6 boards placed across the joists near the chalklines. The blocks should be nailed in flat, centered on the chalkline, to give adequate bearing to the sheathing.

Walk straight across the building, driving two 16d nails through the joist into each block, then turn around and repeat the process going the other way, this time nailing through the joist at a slight

Nailing Edge Blocking

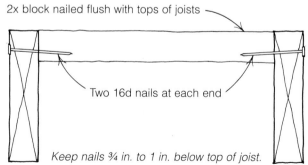

2x block nailed flush with tops of joists

Two 16d nails at each end

Keep nails ¾ in. to 1 in. below top of joist.

angle. Keep the nails ¾ in. to 1 in. below the top edge of the joist so that you won't hit them with a sawblade when you cut the sheathing. After you have finished nailing in all of the regular-length blocks, cut and nail in the specials.

Scattering sheathing

It is more efficient to scatter all the sheets before beginning to nail them in place. One person can easily carry a sheet of ⅝-in. plywood or OSB—the trick is to carry it with one hand underneath and one hand on top for balance, with much of the weight resting against the upper body. Use caution when carrying sheets in windy conditions, as they can act as a sail. Scatter the first row across the joists a few inches in front of the control line, leaving room between the line and the edge of the building to apply an adhesive.

For structural strength

For structural strength, four sheets shouldn't join together at one corner. They should be staggered by beginning every other row with a half-sheet. Staggered sheets add to the horizontal shear strength of the floor, making it structurally stronger.

Installing sheathing

These days many builders fasten down the subfloor to the joists with an adhesive, in addition to nails or even screws. As joists begin to dry out in a house, they can shrink away from the subfloor and cause floors to squeak, which is tough on a teenager trying to sneak in late at night. Squeaks can be greatly reduced, if not eliminated, the entire floor structure strengthened and the quality of the house increased

Sheathing panels should be carried with one hand on top and one on the bottom, with most of the weight resting on the shoulder.

Construction adhesive is applied to joists to bond the sheathing to the floor system.

The first row of sheathing should be aligned with the groove edge on the control line (left side) and the tongue along the edge of the building (right side).

by using a good construction adhesive. To get the best value for your money, buy the large 32-oz. tubes and dispense the adhesive with a caulking gun. One large tube is enough to secure about three sheets of plywood. When you are ready to sheathe, apply a ¼-in. bead on the joists and on any edge blocks that will fall under the first sheet.

If you are using tongue-and-groove sheathing, treat it with care so as not to damage the edges. Damaged edges make it hard to fit two sheets together. Begin laying the first row with the groove right along the control line, hitting the middle of a joist at each end. Tack it down with one 8d nail near each corner to hold the sheathing in position until it can be finished with a hammer or pneumatic nailer. In a dry climate, lay only a few sheets before nailing them off so the glue won't dry out.

In areas of high humidity, leave about ⅛ in. between the ends and edges of the sheets to allow for expansion. This gap can be gauged by eye or by using an 8d nail as a spacer. Sheathing panels sized ⅛ in. smaller in each direction are available in some areas; these allow you to space sheets and still maintain a 4-ft. by 8-ft. module.

If the joist layout isn't totally accurate, some sheets will have to be cut to have bearing on a joist. T&G plywood has to be cut as it is laid down. No measuring tape is needed to mark it to length. Just hold the chalkline in line with the center of the joist and mark the sheet for cutting. Hold one end of the line with your foot, pull it across the sheet, and snap. You can also hold one end of the line on the mark with your little finger of one hand, extend it across the sheet with the other hand and snap the line with the thumb and index finger of the first hand. Once the line is snapped, set the circular-saw blade to the thickness of the sheathing and make the cut.

Let the sheathing run over at the edges of the building. The edges will be snapped with a chalkline and cut at one time after all the floor has been laid. Just be sure to make the cut before nailing so that no nails get in the way.

If you are not using adhesive, it will be quicker to nail off the floor once it is entirely laid. After each row is tacked down, make a mark with pencil or keel on the leading edge of the sheathing to mark the location of every joist. This takes only a moment (it can be done while walking back to begin the next row) and makes it easy for the floor nailer to find the joist.

Once the first row of sheathing has been laid the length of the floor, begin the second row with half a sheet. These 4-ft. pieces can be cut right off the material pile. It's faster to measure and mark half-sheets with a drywaller's T-square than with a tape and chalkline.

Sometimes it takes extra persuasion to unite T&G sheathing. One way to do this is for one person to stand on the sheet, holding it flat. A second person can drive this sheet into place by hitting a 6-ft. long 2x4, laid against the groove, with a sledgehammer, as shown in the photo at right. Don't hit the sheathing directly. Two or three good licks on the 2x4 is usually enough to bring the two sheets together.

Don't nail within 6 in. or so of the leading edge until each succeeding row is laid. A little flexibility on this edge will make it easier to unite the T&G when laying the next course. Sometimes a joist will be bowed at midspan, which does not allow the end of the sheathing to break directly over the joist. Push the joist to the right or left by hand or with your feet until the end of the sheet aligns with the center of the joist and stick a nail through the sheathing into the joist to hold it in place. This saves you from having to make a cut on the sheathing. If the joist can't be moved, a 4-ft. piece of 2x nailed to it can give bearing to the end of the sheet.

As you sheathe up to where the joists lap over a girder, you may need to nail on some 2x scrap pieces to extend the length of one joist and provide a nailing surface. Be sure to mark its location with keel to help keep the nailer on track.

Fitting tongues into grooves sometimes takes extra persuasion. With one person standing on the sheet, another can give it a couple of raps with a sledgehammer. Lay a 2x4 along the edge to protect the groove.

The last row

Many codes call for the last row of sheathing to be at least 24 in. wide because a narrow strip can reduce the shear strength of a floor. For example, if a building is 21 ft. wide and you have laid 5 sheets of sheathing, the last piece would be only 1 ft. wide. If your code won't allow this, snap a chalkline back on the sheathing 2 ft. from the outside of the building then cut along this line with the circular saw. The last row will now be 2 ft. wide and up to code. This row may need to be edge-blocked because there will be no groove to unite with the tongue.

Cutouts

When plumbing pipes are to be installed later there is no need to make any cutouts in the sheathing; sheathe the entire floor and the plumber can drill through it. When the pipes are installed before the sheathing is laid, holes have to be cut in the sheath-ing. The quickest way to make these cuts is to measure from the edges of the sheathing already in place to the center of the pipe and transfer these measurements to the sheet ready to be laid. These holes do not have to be cut out to the exact size of the pipe. Make the cut about ½ in. larger all the way around. So, for a 3-in. pipe, cut out a 4-in. by 4-in. square, 2 in. on each side of the center mark, with four quick plunge cuts (that is, by dropping the spinning blade of your circular saw onto the cut line). With just a little practice it is easy to make this cut by eye. Lift the sheet up and drop it over the pipe.

When sheathing a floor that has a stairwell in it, let the subfloor overhang 2 in. or 3 in. where the stairway will land. This overhang will make part of the last tread on a set of stairs. It can be cut to exact length later when the stairs are being built.

Cutouts in Sheathing

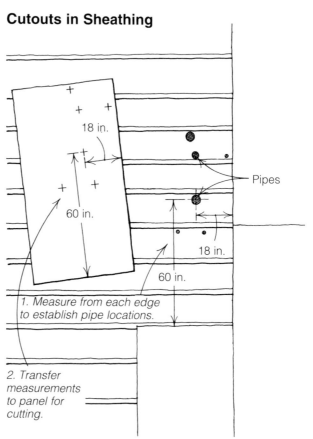

1. Measure from each edge to establish pipe locations.

2. Transfer measurements to panel for cutting.

If your measurements and cuts were accurate, the sheathing should drop right into place.

Nailing sheathing

Building codes usually have a minimum "schedule" (list of requirements) for nailing sheathing (see p. 18), but plans may call for more nails if an engineer determines that they are needed for extra shear bracing. Check carefully to see what size nail is called for and how they are to be spaced. A typical code requirement is 4-6-12, with nails every 4 in. on the perimeter (around the outside), every 6 in. at the joints and 12 in. on center in the field (in the middle of the sheet). The typical floor nail is an 8d box nail, but check the plans in case another type is called for. Occasionally you might be required to use a screw nail or a ring-shank nail, or even drywall screws because of their increased holding power.

Sheathing can either be nailed with a hammer or with a pneumatic nailer. Nailers are certainly much faster, but they are also more expensive to own and operate than a hammer. In addition to the nailer, you need a compressor, air hoses and special nails. Unless you plan to build a number of houses, it's cheaper to use your hammer.

As much as possible, nails need to be driven straight into the center of the joists. If they go in at an angle they can come out the side of the joist and are more likely to cause a squeaky floor later on. Framers call these missed nails "shiners," because

---

SAFETY TIPS

Nailers

Nailers are most often powered by compressed air and, like any tool, they are potentially dangerous. Before using one, make sure that you have been fully instructed in its use and

Pneumatic nailers make it easy to nail down floor sheathing.

care, and read and follow the instruction manual. Don't use a higher air pressure than that recommended by the manufacturer, and always wear eye protection.

When you attach the nailer to an air hose, hold it away from you—nailers will sometimes fire a nail as the air pressure builds up in the cylinder even though they have a double safety. For a nailer to fire normally, you have to pull the trigger and simultaneously push a lever at the nosepiece onto a nailing surface. When you stop nailing, don't walk around with your finger on the trigger—if you bump the nailer against your leg you may release the safety on the nose and fire a nail. If you're just learning how to use a nailer, nail slowly and get the feel of what you are doing.

Take a break from time to time. Straighten up and look around. Nailing on a large subfloor is a little like driving down a long, straight road on a hot day with no air conditioner. You get mesmerized! A nail through the foot will wake you up, just like running off the road, but not without unpleasant consequences.

Nailing Sheathing

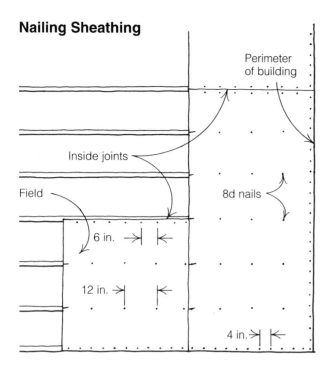

A fast hand nailer needs to learn how to finger out nails one at a time.

you can see them shining alongside the joist from below. Inspectors look for them to judge whether or not a floor has been adequately nailed. A skilled nailer quickly learns to hear and feel when a nail misses the joist and drives another one beside it.

When nailing with a hammer, take a handful of nails from your nail pouch and arrange them so that all the heads are turned up. Finger the first nail out, start it with a light hammer blow, then drive it home with a hard blow or two. But make sure it is started properly before driving it home, because a poorly set nail can fly through the air at an amazing speed if nicked by a hammer blow. That's why you should always wear protective eyeglasses when nailing. As you continue nailing down the joist, try to develop a rhythm with your hands, arms and eyes working together. A good rhythm is the secret of a fast nailer.

When nailing with a pneumatic nailer, you have to pay special attention to crowned joists, which can prevent the sheathing from lying flat on an adjacent straight joist. When you're using a hammer, you can "feel" this situation as the sheathing bounces above the joist. An extra lick with the hammer will bring joist and plywood together. Nailers don't have such sensitivities, although an experienced carpenter will learn to feel for this condition. Be aware that it happens, check the floor for any such gaps and nail any loose sheets down tight.

FRAMING WALLS

3

LAYOUT

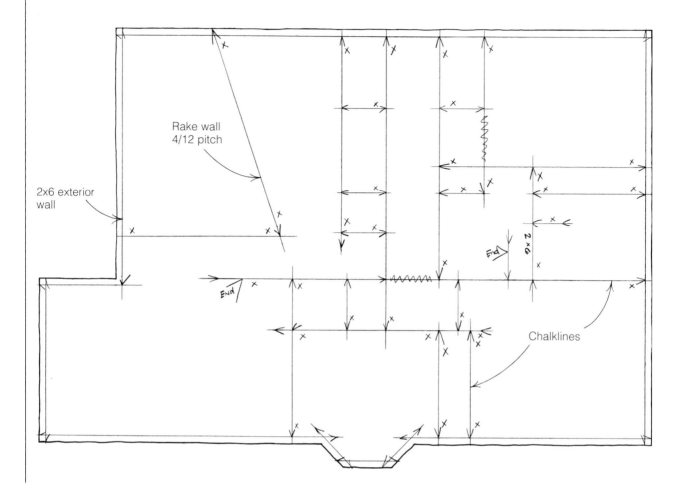

Rake wall
4/12 pitch

2x6 exterior
wall

Chalklines

Wall layout is the process of transferring the floor-plan dimensions to the concrete slab or wooden subfloor of a house, locating exactly where each wall is to be placed. Essentially, it involves taking the architect's floor plan and making a full-sized redrawing on the house floor. In the average house, most layout is quite simple. It consists of marking a series of lines on the floor to form a series of squares and rectangles. These lines represent and locate every wall on the plans. Once the house is framed, the squares and rectangles become bedrooms, bathrooms and kitchens. The floor plan lets the carpenter know the location and size of these rooms.

Some carpenters lay out one wall, then plate it, build it and raise it. Then they do the same with the next wall, one wall at a time. Efficient production framers lay out all the walls first, before going on to the next step. Carpenters don't have to work faster or cheat on quality to increase production, they just need to eliminate wasteful movements. Framing a house is hard work. Good framers make their work easier by thinking about what they are doing and how best to do it.

After the floor-plan lines have been transferred to the house floor, 2x4 or 2x6 plates (i.e., the tops and bottoms of walls) will be cut and placed on all the lines (see pp. 71-78). These plates will tie the framed structure together once they are nailed on.

Before beginning to lay out wall locations, study the plans at home. Look at all the pages and mark them up with colored pens (see pp. 12-13). For example, use red to note the location and size of all posts and beams, green for tall walls, yellow for window sizes. Circle and note everything out of the ordinary, such as walls that need to be framed with different-sized lumber, rake walls and high ceilings. Many buildings these days call for 2x6 exterior walls because they can hold more insulation than the traditional 2x4 wall. Thicker walls may also be required in bathrooms to accommodate all the plumbing pipes used for water and drainage. And when building a three-story building, the first floor needs to be built with 2x6 or 3x4 walls to support the added weight from above.

Layout tools

Wall layout requires very few tools. You will need a 25-ft. and 50-ft. measuring tape, a chalkline, keel (carpenter's crayon) and an awl. Chalklines, which are available in both 50-ft. and 100-ft. lengths, are inexpensive tools, and most framers carry several in their tool bucket. You'll need more than one when you're laying out on damp floors. A good chalkline is geared so that it can be reeled in quickly.

Most framers prefer to use cement coloring rather than chalk when laying down lines. Chalk is much finer than cement coloring and requires you to fill the box frequently. Also, with cement coloring, the line can't be washed away easily by rain, as it can with chalk. In production framing, each job is often done by different people, so the layout crew, for example, will use a different color to chalk their lines than the sheathers so as to avoid confusion.

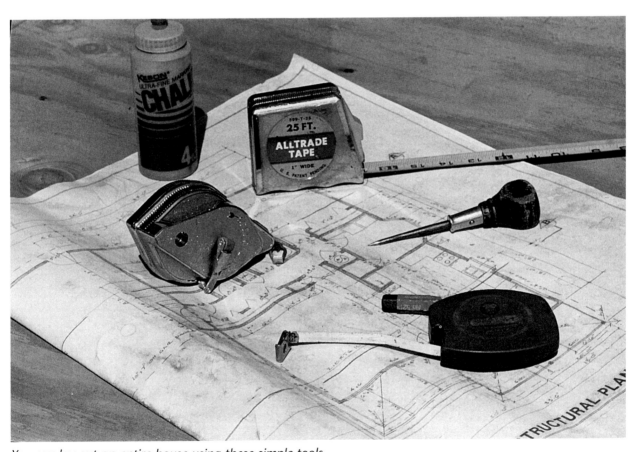

You can lay out an entire house using these simple tools.

If you have to lay out walls that are longer than your chalkline, a long string (dryline) will help. You can stretch the string the full distance of the wall, then make a series of marks along the string to guide your chalkline.

Keel is used for writing on lumber and comes in various colors. Carpenters usually use red or blue because they show up best on lumber. When marking on a freshly poured concrete slab, use white keel because other colors will fade.

A sturdy carpenter's awl driven into a wood floor or a new concrete slab can hold one end of a chalkline or measuring tape. You won't be able to drive an awl into an older, cured slab, in which case a weighted-down coffee can will come in handy for holding the chalkline.

Regular layout

Begin your layout by cleaning up the floor. It's much easier, and safer too, to work on a floor that has been cleaned off. Now take a block the width of the wall (usually 2x4 or 2x6), lay it flush with the outside of the building at each corner and mark on the inside of the block with a carpenter's pencil. If the foundation isn't parallel, make your adjustments at this point before snapping wall lines (see p. 34). Stretch the chalkline tightly from one corner mark to another and give it a snap by pulling straight up; chalklines that are released at an angle can leave a curved line on the floor. Only one chalkline is needed per wall. This chalkline locates the inside edge of the wall.

Once all the exterior wall lines have been snapped, you can work off them to lay out the interior walls. There's nothing complicated about this process. Even though there are many different styles of architecture, the rooms within each style are pret-

An efficient framer will make good use of an awl on a layout job. Stuck in the subfloor, it can hold one end of a chalkline or measuring tape.

The locations for outside wall plates can be marked using a scrap piece of plate stock as a guide. Make sure to hold it flush with the edge before marking. Do this at both ends, then snap a chalkline between the two marks.

ty much the same. A bedroom is still a bedroom. You simply take the dimensions from the plans and transfer them to the floor.

Lay out the longer parallel walls first. Look at the drawing below. On the floor plan, the distance between the outside wall and the parallel bedroom wall is 12 ft. 7 in., measured outside to center. The center of a 2x4 (3½-in.) wall is 1¾ in. from its edge, so this added to the given dimension is 12 ft. 8¾ in. Measure in this distance from the outside of the building and make a mark. Carpenters usually mark locations with an upside-down V (sometimes called a "crow's foot") with the point of the V right on the required dimension. Alongside this mark make a

Laying Out Interior Walls

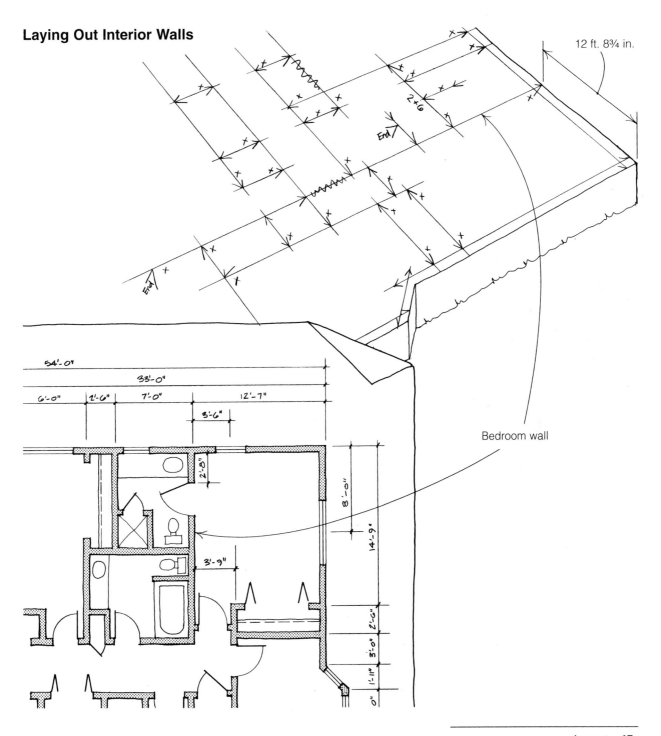

12 ft. 8¾ in.

Bedroom wall

clear X, rather than a simple line, with keel to indicate the location of the plate. This X is important. When it comes time to lay down the 2x plates, the X will indicate on which side of the line to place them. So every time you make a mark for wall location, make an X right alongside it to indicate actual plate location.

Both ends of every interior wall have to be marked by measuring off the outside wall or another known point. Some interior walls will be laid out parallel to other interior walls. Once the end points are located, these marks are connected together with a chalkline to show the location of the bottom plate of a framed wall. Care must be taken to get the correct dimensions from the plan and to be sure whether the measurement is from outside to outside, outside to center or center to center (see p. 14).

It is important to follow plan dimensions carefully, but for most builders, plans are a guide, not a law from which they cannot deviate. For example, a wall between bedrooms might have a pipe coming up through it that has been installed off layout an inch or two. Rather than move the pipe, which is especially difficult on a concrete slab, move the wall a bit and make one bedroom a little larger. This "fudging" is not always possible, of course. For example, exact measurements must be maintained when one wall has to be placed 60¼ in. from another to accommodate a bathtub. If a pipe were installed in the wrong place here, it's the pipe, not the wall, that would have to move.

One problem is that plans don't always have the exact measurements you need. They may show, for example, that the bathroom is 60 in. wide, when in reality it needs to be 60¼ in. or more, so that the tub you are using can be installed with ease. (The tub supplier or manufacturer may be a more reliable source for such information than an architect's plans.) Hallways are another example. A hallway wide enough to accommodate a header and king studs for a 32-in. door needs to be 40 in. between walls.

At times when extra space is needed, walls can be made 1½ in. wide by nailing the studs in flat instead of on edge. This is sometimes done at the back of a closet to add an extra 2 in. in depth.

Very little attention need be paid to door and window locations at this point. All chalklines should be continuous, snapped straight through any wall opening. If you make a mistake in layout, make a clear correction. Rub out an erroneous chalkline with your foot or draw a wavy line through it before snapping another. Anytime you have something out of the ordinary, indicate what is to be done by writing on the floor with keel. For example, if you have a short wall, one that ends in the middle of a room,

The location of interior walls may have to be moved slightly to accommodate plumbing pipes.

indicate that by writing "end" on the layout. If a 2x6 wall is needed in a bathroom in a house that is otherwise framed with 2x4s, write "2x6" right alongside the chalkline (see the drawing on p. 67).

One of the principles of good layout is clarity. On some jobs, the people who do the layout may never see the people who do the plating. Any communication between them must take place in writing on the layout surface. And remember that too many marks can be just as confusing as too few. Keep things as simple yet as clear as possible.

Rake walls

A rake wall follows the slope of the roof. It is necessary when the plans call for a cathedral ceiling. In a rake wall, each stud is a different length. The layout for a rake wall involves drawing a template on the deck, so that the framer can build the wall without having to make any further calculations.

There are two fairly easy ways to lay out a rake wall. The first way is to use a pocket calculator (or your head) to determine the difference in length between the shortest and the longest stud. The shortest stud will normally be the standard wall-stud length (92¼ in. for this calculation). To determine the length of the longest stud it is necessary to know the length of the rake wall. For example, if the building is 33 ft. wide and the rake wall runs up to the center or peak of the roof, then its highest point is midspan, that is, 16½ ft. This figure multiplied by the roof pitch will give you the difference in length in inches between the shortest and longest studs. With a roof pitch of 4-in-12, for example, multiply 16½ ft. by 4 in. for a result of 66 in. Now add this 66 in. to the length of the shortest stud (92¼ in. + 66 in. = 158¼ in.). At midspan on the deck, mark a wall height of 158¼ in. up from the bottom-plate chalkline. Then snap another line between this point and the 92¼-in. mark at the short point and you have a rake wall laid out full scale on the floor. A framer can now come along and cut the intermediate studs to length simply by following the chalklines.

Rake-Wall Layout

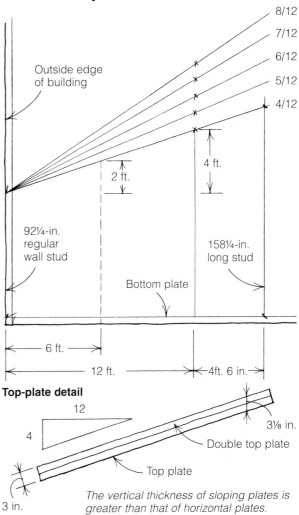

The vertical thickness of sloping plates is greater than that of horizontal plates.

Now, this procedure isn't totally accurate. It overlooks the fact that the top plates, which would account for 3 in. on a level wall, will actually account for slightly more on a rake wall. On a 4-in-12 pitch this would amount to about 3⅛ in. Thus, the studs would have to be shortened about ⅛ in. to maintain perfect accuracy. This difference would be greater on steeper pitches. One way to determine this amount is to lay out the desired pitch, place two pieces of 2x plate along the pitchline and measure their vertical thickness. A much easier procedure is to look under the "spaced 3 in." column under jack rafters in A. F. Riechers' *Full Length Roof Framer* (Box 405, Palo Alto, Calif. 94302). Make sure you have the correct roof pitch.

If the rafters have a bird's mouth, then you may need to shorten the height of the rake walls. To determine how much lower they should be built requires a knowledge of roof-framing theory, which we will discuss later (see pp. 136-142). The easy answer is that they are lowered by an amount equal to the depth of the rafter bird's mouth, generally about 1 in. It is important to note, however, that rake walls normally do not have to be built with total accuracy. Many builders like to have their rake walls 1 in. or 2 in. high for a stick-built roof (see p. 152) and about ½ in. to 1 in. low for a truss roof.

Another way to lay out rake walls was developed by pieceworkers in the days before the pocket calculator was in common use. This method, which eliminates all calculations, is based on the knowledge of how a right triangle works. The trick is to work on a 12-ft. scale and calculate the pitch in feet rather than inches. On our 16½-ft. rake wall, start by marking for the short stud (92¼ in.), as shown in the drawing on p. 69. Next, measure 12 ft. over from this point at a right angle from the edge of the wall. At this 12-ft. point, 92¼ in. up from the wall line, measure up an additional 4 ft. (on a 4-in-12 slope) and make another mark. Snap a line between the short and long point and you have translated the 4-in-12 pitch to the wall layout on the floor. If you extend this line to the 16½-ft. wall, the entire rake wall is laid out. This process works no matter what the pitch of the roof or the length of the wall. On a 6-in-12 pitch roof, you would measure over 12 ft. then up 6 ft.

If there is not enough floor space to lay out for a 12-ft. long wall, cut the wall length and the roof pitch in half. So if the pitch is 4 in 12, you can lay it out as 2 in 6.

The lines snapped on the floor show the length of the shortest and longest studs and the length of the wall. The top line indicates where the top plate will go. Once the bottom and top plates are snapped for the rake wall, don't forget to mark the location of these with an X on the proper side of the line. The bottom plate goes below the bottom line and the top plate above the top line. This way the framer will know exactly where to place the plates and how long to cut the studs.

Rake-wall studs need to be run unbroken from bottom to top plate, because a cathedral ceiling is high and has no ceiling joists coming in at 8 ft. to help stabilize it, especially against wind pressure, as on a standard wall. For walls that are higher than 14 ft., most codes require the walls to be built from 2x6 or double 2x4 studs.

High walls

In a house with an open cathedral ceiling, frequently there will be high interior walls that run at right angles to the rake wall. These walls can branch off at any point to separate one room from another. The height of these walls is determined much like the long stud in a rake wall. Multiply the roof pitch by the overall distance the wall is positioned in from the short point, usually on the exterior wall. If the roof pitch is 4 in 12 and the wall is 10 ft. in, then the studs will be 40 in. longer than the standard 92¼-in. stud.

The plans may show that this wall is a bearing wall for rafters. In this case, build the wall full height. The rafters will require a bird's mouth to provide full bearing on the top plate, and unless the rafter stock is long enough, they will be lapped and blocked directly over the wall. If the rafters need no midspan support then the wall can be built low, the depth of the rafter bird's mouth. Now the rafters can rest on top of the wall without a bird's mouth.

Often the rafters of a cathedral ceiling are supported by a high ridge beam that sets on posts. The length of these posts is figured the same way as high walls: pitch times distance in from the short point. Just remember to subtract the thickness of the beam from the post length plus the depth of the rafter bird's mouth.

PLATING

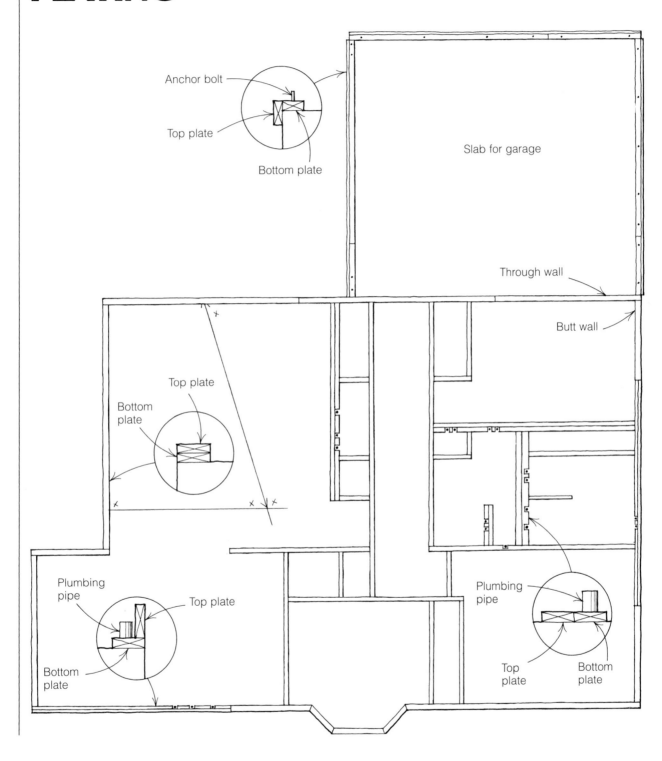

Anchor bolt

Top plate

Bottom plate

Slab for garage

Through wall

Butt wall

Top plate

Bottom plate

Plumbing pipe

Top plate

Bottom plate

Plumbing pipe

Top plate

Bottom plate

Once the layout is complete, it is time to lay down the top and bottom plates. Plates are the horizontal 2x4s or 2x6s that help form the wall frame of the building. When scattering plates, the chalklines are your guide. Two plates are scattered by every chalkline clearly indicating the outline of the rooms, making it easier to visualize the finished structure. The bottom plate will be tacked temporarily to the floor and the top plate tacked to the bottom plate. After all the plates are in position they can be detailed (that is, marked) to show the location of all studs, corners, windows, doors and other framing members (see pp. 89-97).

Plating styles

The speed with which a house can be framed depends a lot on how it is plated. Most platers like to run the longest outside walls through from outside to outside, from corner to corner. These are called "through" or "by" walls. Walls that intersect these through walls, butting into them, are called "butt" walls.

The long outside through walls are plated first, followed by the shorter butt walls. As a general rule, all interior walls that run parallel to the through walls should also run through, from one outside wall to another. Interior walls that parallel butt walls should be plated short to butt the through walls (see the drawing on the facing page). If a house is plated in this manner, it will speed up the framing process considerably because more than one wall can be framed at a time and each wall will be easier to raise into its upright position.

Inexperienced framers will often go around the building laying down plates that run through at one corner and butt the wall at the next corner. This style of plating is called "log-cabining." It is an inefficient method because it makes it more difficult to frame and raise walls in an easy and orderly manner. Rake walls are easier to build if they are framed

Plating involves securing top and bottom plates along their layout lines so that they can be properly marked for the location of studs, corners, windows and doors.

Plating Styles

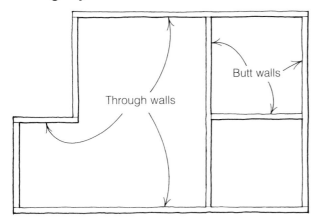

A properly plated house follows a consistent approach to through walls and butt walls. This makes it much easier to frame and raise walls.

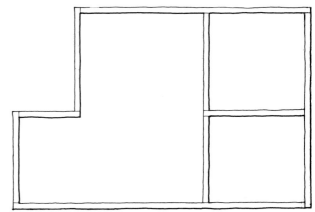

The log-cabin style ignores through and butt walls.

as butt walls, and this may require that other walls be plated log-cabin style, but this is the exception. As a general rule, avoid log-cabin plating.

Scattering plates

For the long walls, it is best to use straight 2x stock at least 16 ft. long so that it can run from room to room without a break and help tie the building together structurally. For structural reasons, most codes require that the ends of top plates be at least 4 ft. from intersecting joints. This is much easier

to do if you use long plate stock. It is important to note that this 4-ft. requirement does not apply to bottom plates, which are fastened directly to the floor and are not subject to the same type of stress that may affect the top plates. (Top and bottom plates do not have to break in the same place.) Further, using straight stock now, especially for the top plate, means that it will be easier to ensure that the walls themselves will be straight once all the framing is completed.

Scatter all of the plates before cutting them to length and tacking them in place.

Check the plans or look at the layout to determine whether the exterior walls are 2x6 or 2x4 (most interior walls are 2x4) and position a top and a bottom plate next to every line. If you are building on a slab, it will be easier to mark and drill the holes in the bottom plate and drop the plates over the anchor bolts before scattering the rest of the top-plate stock (see p. 78). Many interior walls are short and won't require long pieces of plate. The important point is to carry and position as many plates as possible before beginning to cut and nail them in position.

Positioning and securing plates

With the plates scattered, begin by plating the long outside through walls. Place a piece of stock flat down on the floor, flush with the chalkline on the outside edge of the building. Remember that the ends of through plates must extend beyond any butt walls. Gauge this distance with a short piece of plate stock or a measuring tape. Run both plates continuously, ignoring door and window openings. The bottom plate in doorways will be cut out later when the jambs are set.

In an average house, many feet of plate can usually be tacked in place before any have to be cut to length. Saw cuts need to be made at the end of a wall, around pipes and to ensure a 4-ft. lap of the

Begin by plating the outside walls.

Plates are aligned with the chalkline and then tacked to the floor temporarily with 8d nails.

top plate at intersecting walls. Tack down as many plates as you can before picking up the saw. Tacking is done by holding the plate on the X side of the chalkline and driving an 8d nail about 1 ft. from each end of the plate and one in the middle. (Remember, these are only temporary nails—you don't have to use 16d nails.) Keep nails away from corners and channels where other walls will intersect. Next, put a top plate flat on the bottom plate and tack the two plates together with two or three more nails. These nails will hold the plates secure until they are detailed. Continue plating the full length of the wall, making sure that no gaps are left where one plate butts another.

Sometimes you can't position the bottom plate flush with the chalkline because plumbing pipes are coming up through the floor. When this is the case, you will have to cut notches in the bottom plate, leaving about ½-in. clearance around each pipe, be-

fore tacking it to the deck. The top plate doesn't need to be notched, but it will have to be tacked on to the bottom plate differently. On exterior walls it can be held over the edge of the floor and tacked to the side of the bottom plate, or it can be placed on edge on top of the bottom plate and alongside the pipes (see the details in the drawing on p. 71). On interior walls the bottom plate is notched around the pipes, tacked to the line, and the top plate is simply laid flat alongside it, allowing both plates to be marked for framing at the same time.

Cutting plates

With a little practice, you should be able to cut plates by eye. All plates need to be cut accurately, but you should take special care to cut the top plate as close to perfect as you can. If this plate is cut precisely, then it will be easy to make sure that the house, and every room within it, is square and straight.

The bottom plate needs to be cut or notched to fit around plumbing pipes.

Cutting plates to length can be done by eye. Use the chalkline or plates that have already been tacked down as your visual guide.

Supporting Lumber During a Cut

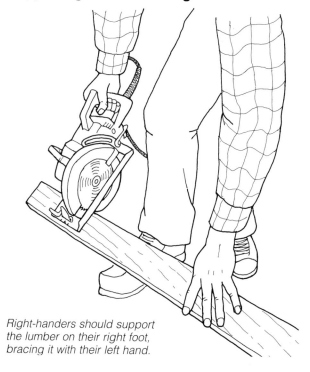

Right-handers should support the lumber on their right foot, bracing it with their left hand.

Cut plates by sighting down to the chalkline and cutting them to length straight across. It is especially easy to cut plates where butt walls intersect through walls. The butt-wall plate will be too long and will have to lay over the top of the through wall and be cut flush. After a few tries, and a few mistakes, you should be able to cut as accurately as if you were measuring and marking each plate with a tape and a square. When working with a partner, it is a time-saver to have one person do all the cutting and the other all the positioning and tacking of plates.

When a plate can't be cut to length by eye, framers like to use another method. Mark the piece to length, then brace it across your foot, as shown in the drawing at left, and make the cut. There's really no need to use a sawhorse. The lumber is held stationary with one foot and one hand, allowing the cut-off end to drop free without binding the saw.

Butt the intersecting wall plates gently against the through walls so as not to knock them off their lines, cut them to length and tack them in place. Try to keep all through walls parallel with the long outside walls. Watch for interior walls with pipes in them that need 2x6 plates. When layout marks on the floor indicate that a wall is 1½ in. wide, rip a 2x4 to 1½ in. Marking the rip line can be done by simply holding a pencil 1½ in. in on the board, striking a line the length of it using your fingers as a guide, and then cutting with the saw. The worm-driven saw is actually set up to rip 1½ in. wide, that being the distance from the left side of the table to the blade. To rip this width hold the left side of the saw table to the outside of the 2x stock and make the cut.

Most plating is not complicated but is mainly a matter of using common sense. Learn to organize plating into a system of through and butt walls. Once you do, you should be able to tackle almost any house design.

Plating a bay window

Inexperienced carpenters often plate for bay windows that extend from the house at 45° by cutting one plate square and fitting the adjoining plate to it at a 45° angle. But it is easier to frame these walls when each plate is cut at approximately 22½° to form the corners. This can be done without the use of a square or angle divider of any kind. Simply lay the plates in position right on the chalklines. Let the two plates forming the main exterior wall run past the bay window opening about 1 ft. Do the same for the plate that will form the front of the bay. Now

take two shorter pieces of plate that will form the sides and place them on the line and on top of the other two plates. With the saw, make a cut from corner to corner across these short plates. This gives you about a 22½° angle on the side plates and will leave a saw mark on the plates below indicating where they need to be cut. Once all the bottom plates are cut to length, they can be used as patterns for the top plates.

Plating a rake wall

A rake wall is plated the same as any wall with a top and bottom plate. It is usually best to butt a rake wall into a through wall. This way it can be easily framed, raised into position and tied into intersecting walls. Rake walls are often the last to be built and raised. If they were allowed to run through, it would be difficult to lift them into an upright position because the butt walls would be in the way.

The top plate on a rake wall runs up at an angle, so it will be longer than the bottom plate. But, for now, cut a top plate the same length as the bottom plate and position both plates flush with the chalkline on the outside wall. This short top plate will be used as a layout guide to align the top ends of the rake-wall studs (see p. 104). Framers normally cut the second, longer top plate when they are actually building the wall.

Plating on a slab

When plating on a concrete slab, you usually have to contend with bolts embedded in the slab. Bottom plates have to be drilled so that they can fit over the bolts and be secured to the slab. On an outside wall the top plate can be held over the edge of the floor and tacked to the side of the bottom plate, or it can be placed on edge on top of the bottom plate. Since the bottom plate will be in contact with concrete, it is necessary to use pressure-treated lumber (see pp. 28-34).

Marking the location of bolt holes on bottom plates is fairly easy with the help of a bolt-hole marker. With this simple tool there is no need to go through the slow process of marking bolt location with a measuring tape and a try square. A bolt-hole marker transfers the location of the bolt to the plate. For years, framers built their own bolt markers (see

Plating a Bay Window

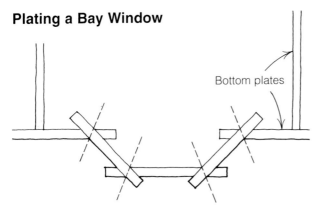

Bottom plates

Lay the two side plates of the bay window on top of the other wall plates. Then cut the two from corner to corner.

The bottom plate on a slab floor can be marked for the location of anchor-bolt holes using a bolt-hole marker.

Making a bolt-hole marker

Take an 18-in. by 1½-in. by ⅛-in. metal plate strap. Cut a notch in one end of the strap so that it will fit around a ½-in. bolt. From the center of this notch measure back on the plate strap 3½ in. for a 2x4 plate and 5½-in. for a 2x6 plate. Drill holes at these points and put in ³⁄₁₆-in. by ¾-in. stove bolts. Put a bend in the plate strap to make it easier to use.

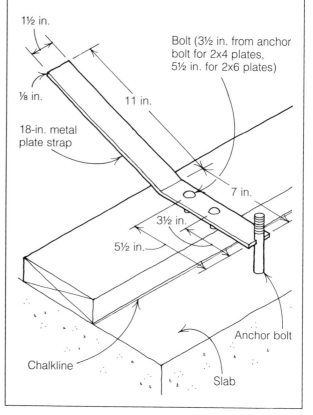

the sidebar at right), but they can now also be purchased commercially (Pairis Enterprise, 2151 Maple Privado, Ontario, CA 91761).

To use the bolt-hole marker, place the plate on the inside of the chalkline opposite to where it actually goes. Check to make sure that the plate is properly positioned directly on the line and end for end. Hold the bolt marker against the bolt, straight out over the plate. Tap the ³⁄₁₆-in. stove bolt on the bolt marker with a hammer to leave a mark on the wood for drilling. Drill the holes and drop the plates over the bolts. The plate should sit in place right on the line, where the top plate can then be tacked to it.

Once framed and raised, interior walls are normally secured to concrete floors with a metal pin (see p. 114). For now, the plates have to be held in place temporarily, just as on a wooden deck, until all plating is finished and the plates can be detailed in preparation for framing. If the slab was poured recently, plates can often be tacked down to the concrete with an 8d nail driven straight through them into the concrete. If the slab has cured and is hard, the plates can only be laid into position. The bottom plates of intersecting walls can be toenailed to the exterior walls and to one another. The top plates are then cut to length and tacked to the bottoms. Because these walls may not be attached as securely to the floor as on a wooden deck, they have to be treated a bit more gently so that they stay in place until all detailing is complete. If they happen to get knocked off the chalkline, errors in marking the location of intersecting walls can occur.

HEADERS, CRIPPLES, TRIMMERS AND ROUGH SILLS

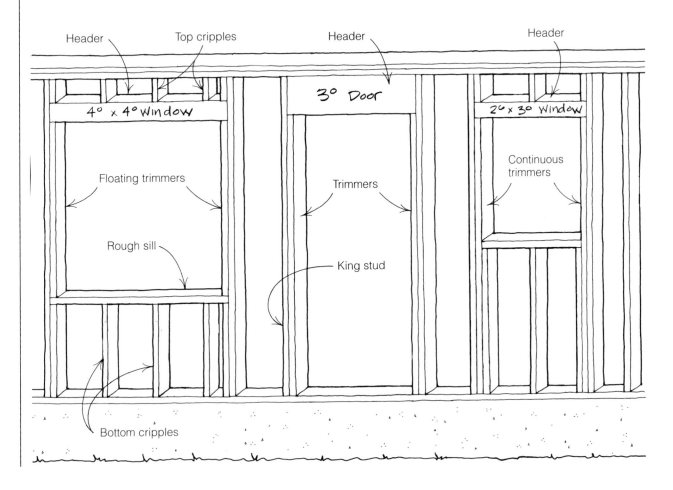

Header · Top cripples · Header · Header
4⁰ x 4⁰ Window · 3⁰ Door · 2⁰ x 3⁰ Window
Floating trimmers · Trimmers · Continuous trimmers
Rough sill · King stud
Bottom cripples

Headers (also called lintels) run horizontally over the tops of door and window openings. They distribute the weight from above and permit an opening below. The header itself is supported by vertical trimmers. Cripples (also called jack studs) are short vertical studs that connect the headers and the top plate ("top cripples") or the rough window sill and the bottom plate ("bottom cripples"). Some people object to calling these short studs "cripples," for obvious reasons, but the name persists. All headers, and the cripples and trimmers that support them, have to be counted and cut to length.

A standard height of door and window headers is 6 ft. 10 in. from the floor to the bottom of the header. This dimension can vary regionally, so check your plans to determine the correct header height. There are some exceptions: Pocket sliding doors and some closet bifolds require the header to be 2 in. higher to leave room for the tracks that carry the door. Garage-door headers are also usually installed at 7 ft. Sliding-patio-door heights vary according to brand. Some may need to be lower than 6 ft. 10 in. Top cripples are cut to bring headers down to the required height. A rough-framed window opening in a wall has a sill that runs parallel to the plates. Bottom cripples are cut to bring the sill up to the correct height.

Types of Headers

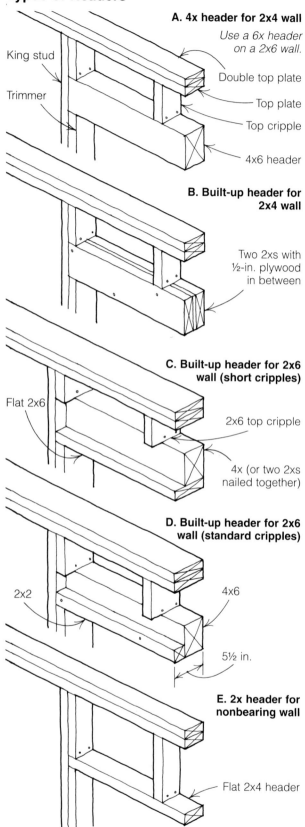

A. 4x header for 2x4 wall

Use a 6x header on a 2x6 wall.

King stud

Trimmer

Double top plate

Top plate

Top cripple

4x6 header

B. Built-up header for 2x4 wall

Two 2xs with ½-in. plywood in between

C. Built-up header for 2x6 wall (short cripples)

Flat 2x6

2x6 top cripple

4x (or two 2xs nailed together)

D. Built-up header for 2x6 wall (standard cripples)

2x2

4x6

5½ in.

E. 2x header for nonbearing wall

Flat 2x4 header

Doors and windows

The number and sizes of door and window openings needed for a structure can be found by studying the floor or framing plan (see p. 16) or the door and window schedule (see p. 18). This is the kind of work that can usually be done at home. On residential plans a door size might be shown as $3^0\ 6^8$, which means that the actual door going into this opening will be 36 in. wide (3 ft. 0 in.) and 80 in. tall (6 ft. 8 in.). (Note that many codes now require all passage doors to be at least 32 in. wide.) A standard window size might be shown as a 2^6 by 4^0. The first number always notes the width of the window sash (30 in.), the second number notes the height (48 in.). Carpenters talk about openings in this way: "two-six by four-oh."

It is important to remember that this "call-out" size, as it is called, refers to the actual size of the door or window and not to the jamb or frame that holds it. Sometimes only the actual size of doors and windows is given on the plans, but framers can still use this information to determine the header length, the rough-opening width (the distance between trimmers) and the rough-opening height (the distance from the header to the rough window sill or floor).

Headers

Headers can be cut from 4x stock for 2x4 walls (drawing A at left) or 6x stock, when available, for 2x6 walls. Built-up headers, made from two pieces of 2x stock with ½-in. plywood sandwiched between and nailed together with 16d nails at 16 in. on center, work well for 2x4 walls (drawing B). Nail a 2½-in. ripping between 2x stock for 2x6 walls.

For small openings, like a 3^0 door in a 2x6 wall, headers can be made from 4x6s placed flat. Alternatively, a 4x header can be placed on edge to provide added strength, with a 2x6 nailed flat on the bottom edge to bring the 4x out to 6x width (drawing C). If you use this flat 2x, cripples will need to be cut 1½ in. short. Nail the 2x6 cripples to the 4x and then to the 2x6 top plate. Yet another method allows you to use the standard-length cripple. If the header is a 4x6, nail on the 6¼-in. cripples. On the bottom edge of the header, nail on a full 2-in. wide ripping to make the 3½-in. header 5½ in. wide (drawing D).

Either method will give adequate backing for drywall or siding.

On nonbearing walls, that is, walls that support no weight from above, it is usually permissible to use a 2x laid flat for a header (drawing E). But using a flat 2x means that you have to cut longer cripples and make sure they are placed only in nonbearing walls.

Increasingly, builders are using different varieties of "engineered wood" products in various parts of the house frame (see p. 22). Laminated-veneer lumber, for example, makes very good headers and beams because it is not inclined to shrink or twist.

The standard length of a header for doors and wooden windows is 5 in. over the call-out size. Thus the header for a 3^0 opening will normally be 41 in. The extra 5 in. leaves room for a 1½-in. trimmer and a ¾-in. jamb on each side plus ½ in. to plumb the jamb. If prehung doors are being used, builders prefer to cut the headers 5½ in. longer than call-out size. The added ½ in. leaves extra room to insert the frame.

Rough-opening sizes for aluminum and vinyl-clad window frames are becoming standardized. Before cutting headers or cripples for these windows, check and double-check with the supplier or manufacturer for the recommended rough-opening sizes. The thickness of the material covering the jambs (for example, ½-in. vs. ⅝-in. drywall) also makes a difference in header length. Once the rough opening is determined, add 3 in. to the header length to allow for a 2x trimmer on each side. For example, a 36-in. rough window opening takes a 39-in. header. Metal windows don't have jambs like wooden windows, so the wall covering covers the trimmer and butts into the window. The rough-opening size usually leaves ¼ in. between the trimmer and the window so that the drywall covers ¼ in. of the window jamb. For ⅝-in. drywall, cut the header ¼ in. longer and leave ⅜ in. between the trimmer and window. Just keep in mind that there are a lot of variables in door and window sizes, and a framer needs to know exactly what he or she is doing. When in doubt, check with someone who knows.

Typical Door Frame

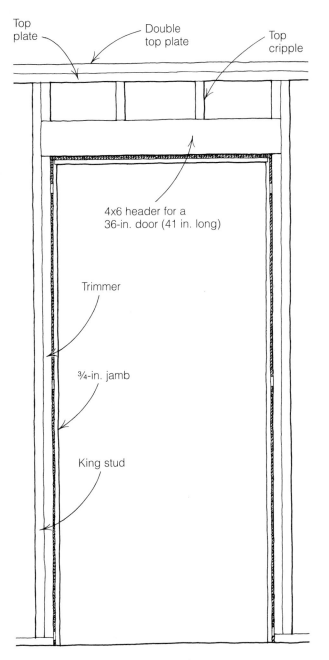

For a standard 3^0 door the header is cut 5 in. long to accommodate a 2x trimmer and a ¾-in. jamb on each side. The extra ½ in. leaves room to plumb the jamb.

The plans may call for double 2x trimmers under each end of a long header, such as for a garage door. If double trimmers are specified, simply add 3 in. to the header length.

The width and thickness of headers are usually given on the floor or framing plan or in a plan detail showing structural requirements. Header size can also be determined by referring to the local building code or by talking to the architect. A general rule of thumb for determining header sizes for a one-story building is shown in the chart below. Some builders like to standardize the building process by using 4x6s for all openings up to 6 ft. This general rule may not apply if there is an extra load above, such as a second story or a balcony cantilevered beyond the opening. Many codes require at least a 4x12 or 4x14 header for a double garage door.

Occasionally the plans will call for two windows to meet in the corner to capture a good view. When this occurs on a 2x4 wall, add 2 in. to the overall length of each header and then miter the outside corner at 45°. The header is lengthened 2 in. because a 4x4 instead of a 2x4 will be used in the corner as the supporting trimmer. The miter can be made with a beamsaw or by cutting both ways with a smaller circular saw. Cut the 4x4 trimmer full length so it can be nailed directly to the bottom plate rather than to the rough sill. Simply cut the rough sills 1½ in. short to allow room for the trimmer to go from header to plate. The same can be done on a 2x6 wall, using a 6x6 corner post.

Header over Corner Windows

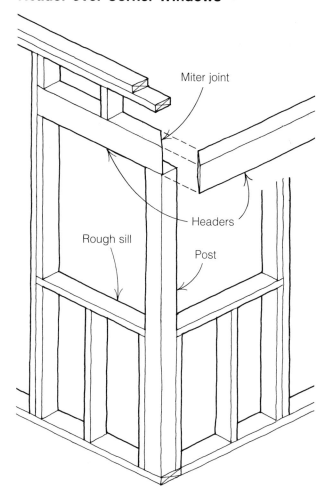

Corner window headers can be mitered to sit on a 4x4 or 6x6 post (trimmer) that extends to the bottom plate.

SIZING HEADERS	
Maximum length of header	**Header size**
4 ft.	4x4
6 ft.	4x6
8 ft.	4x8
10 ft.	4x10
10 ft. to 16 ft.	4x12

Solid-stock headers

In houses using 92¼-in. studs with a 6-ft. 10-in. header height, all top cripples can often be eliminated by installing 4x12s for both window and door headers. A 4x12 is 11½ in. high. By nailing it hard up against the top plate the header height will be dropped to 6 ft. 10¼ in., close enough for a 6-ft. 10-in. opening. Further, nailing headers directly to the top plate strengthens the frame. When a header has to be 2 in. higher for pocket sliders and bifolds, a 4x10 nailed to the top plate works well.

Solid-stock headers like this 4x12 can be nailed directly to the top plate, thereby eliminating all top cripples.

You can also use a 4x6 or a 4x4 for a door header when conventional door jambs are being used. This header is also nailed hard up against the top plate, requiring no cripples. When setting the jamb, trimmers are cut and installed full length to the header. Once the jamb is set, the cut-out section of the bottom plate is nailed on top of the jamb head for drywall backing. When using prehung doors, the header needs to be set at 6 ft. 10 in. so that drywall can be attached before the prehung unit is installed.

There are times when using solid-stock headers won't work. Frequently, when a house has a forced-air unit, the hallway ceiling will be lowered to make room for heating ducts. These ducts lead to the various rooms in the house and need a space above the header through which hot or cool air can be vented. Headers with cripples leave adequate space for these vents.

Whenever you have doubts about the proper window header and cripple size, leave the opening an extra 1½ in. or 2 in. larger. Once the window frame has been delivered and the exact size is known for sure, adjustments can be made as necessary. It is always easier to make a hole smaller than it is to make it larger.

jim haun construction, inc
FRAMING CONTRACTORS
6846 VALJEAN AVENUE — VAN NUYS, CALIFORNIA 91406
780-1919

CUTTING LIST FOR __Peterson House__ ADDRESS __Coos Bay, Ore.__ PHONE _____

JOB ADDRESS _____

SYMBOL	OPENING SIZE	NO. OF PIECES	SIZE	LENGTH	TRIMMERS		CRIPPLES	
					SIZE	LENGTH	TOP	BOTTOM
			DOORS					
	2⁸ X 6⁸	1	4X4	29"	2X4	80½"	8¼'	
	2⁸	10	ς	35'	ς		ς	
	3⁰	1	4X8	41"	2X6	(	4¼"	
	5⁰	3	4X6	65"	2X4	(	6¼"	
	7⁰	1	(	89"	(		ς	
	8⁰	1	ς	101"	)		ς	
			WINDOWS					
	2⁰ X 4⁸	2	4X4	27"	2X6	48"	8¼'	31"
	2⁰ X 3⁸	2	ς	33'		36"	ς	43"
	3⁰ X 3⁸	2		39"	(	ς	ς	ς
	4⁰ X 3⁸	1	4X6	51"	)	)	6¼"	)
	3⁰ X 4⁸	2	ς	39'		48"	ς	32⅜"
	6⁰ X 4⁸	8	ς	75'	)	ς	)	ς
TOP CRIPPLES		35	2X4	6¼'				
		45	2X6	ς				
		15	ς	8¼"				
		43	2X4	ς				
BOTTOM CRIPPLES		10	2X4	31"				
		50	2X6	ς				
		8	ς	43"				
		8	2X4	ς				

This list is supplied as a service only · No responsibility is assumed.

Cutting List for Headers, Trimmers and Cripples

Cutting list

The next step is to count all the headers that will be needed. Go through the floor plan and note down on a list the number of 3⁰ doors that are required (see the chart above). Once an opening is counted, check it off on the plans so it won't be counted again. Then look for the 2⁸ doors, and so on, until all door and window openings have been accounted for, marked down on the cutting list and checked off on the plans.

Once the headers have been listed, you can quickly estimate the number of top and bottom cripples needed. Headers and rough sills need a cripple stud

The story pole measures from the subfloor to the bottom of the top plate. The distance from the top of this 4x4 header to the top of the story pole equals the length of the top cripples.

Making a story pole

Tack a piece of 2x stock to the bottom of a wall stud. Measure up from this bottom plate the standard height of headers (usually 6 ft. 10 in.) and make a pencil mark. From this mark measure on up the story pole for the header size. If the header is a 4x6, measure up 5½ in. What is left over on the story pole, 6¼ in., is the length of the top cripple studs for a 4x6 header. If the header is a 4x4 the top cripples will be 2 in. longer; if it is a 4x8 they will be 2 in. shorter.

The length of bottom cripples under the rough sill is calculated in much the same way. For wooden windows you need a rough opening height of 3 in. over the call-out size. So for a 3⁰ window measure down from the bottom side of the header 36 in., plus 3 in. for jamb head and sill, and make a pencil mark. Next make another mark 1½ in. lower to allow for the rough sill. What is left over on the stud, 40 in., is the length of the bottom cripple studs for a 3⁰ wood window. For a 4⁰ window subtract 12 in. from this figure, for a 5⁰ window subtract 24 in., and so on. Some codes call for a double sill for windows over 6 ft. wide. In this case, subtract another 1½ in. from the length of the bottom cripple.

The bottom cripples for metal or vinyl-clad windows are calculated the same way. Make sure you have the correct rough-opening size from the supplier for the windows you plan to use, measure down from the header this distance plus 1½ in. more for the rough sill, and what remains on the story pole is the length of the bottom cripples. If the builder wants to install a wooden sill in the finish, more room may need to be left in the opening, especially if the metal frame will sit on top of the finish sill.

6¼ in. (top-cripple length)

4x6 header

92¼ in. (stud length)

6 ft. 10 in. (header height)

39 in. (window-trimmer length; rough opening for a 3⁰ wood-frame window)

Rough sill

40 in. (bottom-cripple length)

Bottom plate

on each end and one every 16 in. So for a 3⁰ header cut four cripples, for a 4⁰ header cut five, for a 5⁰ cut six, and so on. It doesn't hurt to cut a few extras because cripples sometimes split when they are being nailed on.

The length of top and bottom cripples and trimmers is easy to determine using a story pole. This is a simple tool, made from a wall stud in this case, marked to show different header sizes and the height and length of door and window openings in a framed wall (see the sidebar above). Using the story pole you can see, for example, that openings with 4x6 headers need 6¼-in. top cripples to bring them down to the standard 6-ft. 10-in. height. Note the length and number of cripples needed for every opening on your cutting list. The length of window trimmers, the rough-opening height, is the distance from the header to the rough sill. The length of door trimmers is the distance from the header to the bottom plate.

Cutting headers, cripples and rough sills

On some jobs, once a cutting list has been made it may be cost-effective to have all the headers and cripples precut at the lumberyard. Cripples can be cut from #3 stock, which is a less costly grade of lumber. With gang saws that make multiple cuts, this can be done quickly and inexpensively. You may even find a lumber supplier who will do the cutting for no extra charge just to get your business.

If you are doing the cutting yourself, begin by gathering up all the scrap pieces of 2x, 4x and 6x stock. A fast way to cut the headers and cripples is with a 16-in. circular saw ("beamsaw"). A 10-in. or larger power miter saw ("chopsaw") or a radial-arm saw can also be used.

If these tools aren't available, you can use a regular circular saw. One way to speed up production is to set the cripple stock on edge and shove each piece against a plated wall. Say you need 30 pieces of 2x4 cut at 40 in. for bottom cripples. Make sure that you have 30 pieces lined up along the plate, then measure up 40 in. on each end piece and snap a chalk-

line across them. Cut along the chalkline and then pick up each piece individually to finish the cut.

As a general rule, cut the longest headers and cripples first. It is a bit embarrassing to cut all the 3⁰ headers first, for example, only to find that the remaining stock is too short for an 8⁰ patio window. When the long headers and cripples are cut first, the cut-off ends can be used for the shorter ones.

Work your way down the cutting list, making all the cuts. Windows need both a header and a 2x rough sill. Cut both the same length and at the same time, 2x4 sills for 2x4 walls and 2x6 sills for 2x6 walls, and tack the sill to the header with an 8d nail. When using aluminum-frame windows, the sill can be the same length as the header or it can be cut 3 in. short to allow for full-length precut trimmers. For wooden windows it is easier to use a full-length sill with floating trimmers that are cut during the framing process.

Write on the sills or headers with a piece of keel the door and window sizes and the top and bottom cripple lengths. For example, when you cut a 4x6 door header 41 in. long, write on it "3⁰" along with "6¼-in. top crips" (cripples). Having the size written on the header makes it easy to locate it on the plates simply by matching its size with what is written on the floor plan. Having the cripple lengths noted on the header makes it easy to scatter the right lengths

Gang-Cutting Cripples

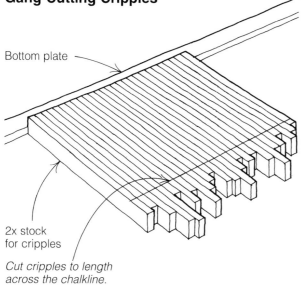

Bottom plate

2x stock
for cripples

Cut cripples to length across the chalkline.

Marking Headers and Rough Sills

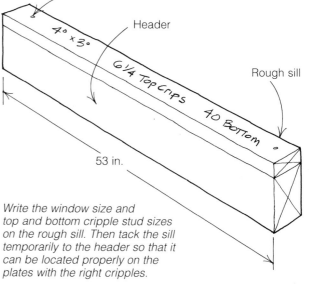

8d nail

Header

4⁰ x 3⁰

6¼ Top Crips

40 Bottom

Rough sill

53 in.

Write the window size and top and bottom cripple stud sizes on the rough sill. Then tack the sill temporarily to the header so that it can be located properly on the plates with the right cripples.

The preferred tools for cutting headers, cripples, rough sills and trimmers are a 16-in. circular saw (above left), a power miter saw (above right) or a radial-arm saw (left).

Eric Haun

make a mark on the plates (see the top drawing on the facing page). Then measure to the center of the header, make another mark and match this mark with the one on the plates. When no measurement is given, use an architect's scale or a measuring tape to calculate the location of a header.

When an opening is shown to fall near the corner of a room, locate the header on the plates one stud (1½ in.) out from the corner. This stud will be the king stud and will nail on to the end of the header. Once the wall is framed and the trimmer nailed under the header this will leave 3 in. of wall space in the corner to nail on drywall and door or window trim.

With a little practice, headers that center in short walls can be spotted by eye. Set the header on the plate and center it by seeing, or measuring, that there is an equal amount of space on each end.

Headers that center on longer walls can still be spotted quickly. Simply shove the header into one corner of the room parallel to the wall it belongs in. Measure the amount of the wall not covered by the header. Divide this measurement in half and measure out from the corner this distance. Placing one end of the header on this mark will center the header on the wall. This kind of "trick of the trade" allows you to accomplish more with less work.

Picking up and carrying headers with a straight-claw hammer can save wear and tear on your back.

Place each header on the plates, near its location in the wall frame.

Spotting Headers

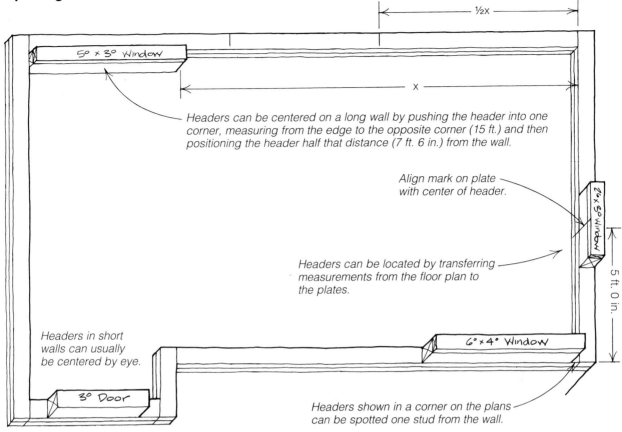

Headers can be centered on a long wall by pushing the header into one corner, measuring from the edge to the opposite corner (15 ft.) and then positioning the header half that distance (7 ft. 6 in.) from the wall.

Align mark on plate with center of header.

Headers can be located by transferring measurements from the floor plan to the plates.

Headers in short walls can usually be centered by eye.

Headers shown in a corner on the plans can be spotted one stud from the wall.

Marking header location

Methods of marking header locations differ regionally. The rule of thumb is to keep detail marking as simple as possible. Mark the location of the end of the header, and leave an X beside the mark for the king stud and a line on the other side to show where the window or door opening will be. There is no need to put a T to show the location of the trimmer or a C for the end cripple. Trimmer and cripple location are self-evident to an accomplished framer; beginning carpenters need only be shown that trimmers go under the ends of each header and end cripples need to be nailed on the ends of each header and rough sill. Their location doesn't need to be marked again and again. Putting unnecessary marks on the plates is time-consuming and usually more confusing than helpful.

Once the header is spotted, note its location on the plates with a piece of keel. Just mark down from each end of the header across both plates. Usually

Marking Header Location on Plates

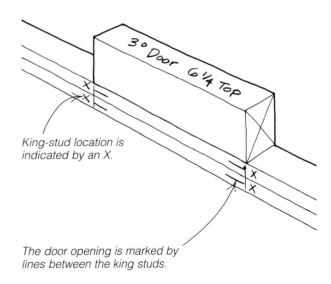

King-stud location is indicated by an X.

The door opening is marked by lines between the king studs.

there is no need to use a square; with practice you can mark accurately across two plates by hand. On outside walls make these marks on the outside; on interior walls, on the side the stud layout will be. Next to this end line mark an X on both plates on the side away from the header to indicate the king-stud location. On the other side of the line make a longer, straight line along both plates. This line indicates that there will be an opening at this location and no studs should be nailed in here.

That's all there is to it. These marks are important and need to be made simply, clearly and accurately. Plates will be pulled up and moved around during framing. Unclear or inaccurate marks mean that some framing members may be nailed in incorrectly and have to be changed later.

Marking corners and channels

At the point where walls intersect, extra studs are required to provide adequate backing to nail the walls together and attach drywall. When these extra studs are needed at a corner, the two- or three-stud backing is called a corner. When partition walls intersect midspan, the three-stud backing is called a channel. Corner and channel locations need to be detailed accurately. These marks will be used by framers to nail in the studs and make cuts in the double plate. Accurate detailing will make the walls easy to plumb and line once they are raised (see pp. 116-124). To detail corners and channels you need a pencil, keel and a marking tool, which can be pur-

Making a channel marker

Take two pieces of 2x stock, 10 in. to 12 in. long. Turn one piece up on end and place the second piece flat on it to form a T. With the second piece protruding over the first about 3 in., nail the two together with 16d nails.

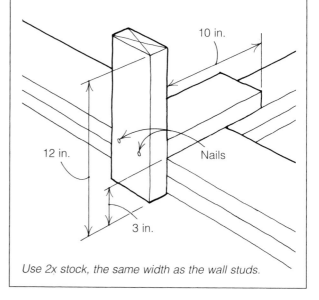

Use 2x stock, the same width as the wall studs.

chased (Pairis Enterprises, 2151 Maple Privado, Ontario, CA 91761) or made following the directions in the sidebar above.

Corners and channels need to be marked on the inside, the top and the outside of the plates, as shown in the drawing at left. Where a wall butts a through wall, place the channel marker on the butt wall, over the through wall, and let it run down 3 in. over the plates on the outside. Mark along all sides and the top with a carpenter's pencil. Now, no matter which way the plates are turned during framing, the marks will be there showing exactly where to nail the intersecting walls and where to cut out the double plate so another wall can tie in with a lap. When walls intersect at different heights, plates can't lap so the double plate doesn't get cut. Mark PT with keel at this point to indicate that the double plate

Marking Corners and Channels

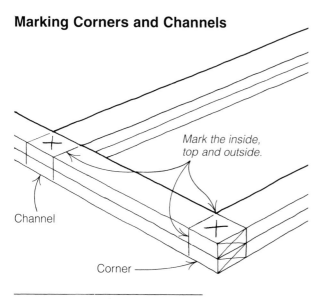

Mark the inside, top and outside.

Channel

Corner

A channel marker allows for quick and accurate detailing of corners and channels.

is to be "plated through." On the top plate of the through wall many carpenters will make an X or an O with keel at the point where the walls intersect.

Detailing the specials

Besides doors and windows, the plans must be checked for the location of medicine cabinets, bathtubs and showers, wall heaters, air-conditioning units, posts under beams, clipped (shortened) or tall walls and stud lengths that vary from the standard. Sometimes detailing will be needed to position built-ins such as an ironing board, a recessed cabinet or a place to set a phone. These locations must be indicated on the plates so that they can be allowed for during the framing process. It is much more efficient to nail in everything that needs to be built into a wall while it is being framed flat on the deck. After the wall has been raised, it takes much longer to add an opening or even nail in a single stud.

An important point for carpenters to remember is that the wood frame has to accommodate plumbing, heating and electrical features. Effort must be made, either by reviewing plans, from experience or

Detailing Bathroom Specials

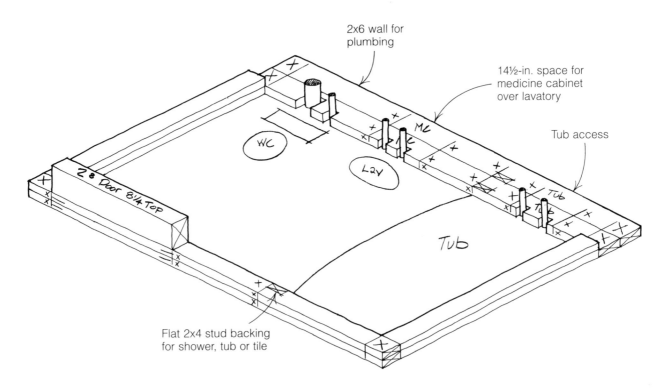

2x6 wall for plumbing

14½-in. space for medicine cabinet over lavatory

Tub access

Flat 2x4 stud backing for shower, tub or tile

through talking to other tradespeople, to make sure that the wall is framed so that others can do their work without having to remodel the structure. Attention paid to proper detailing at this point can save time for all later on.

The standard medicine cabinet (MC) is about 14 in. wide and fits between two studs located at 16 in. on center. Plans often show medicine cabinets centering over a lavatory. If no measurement is given, scale the cabinet's position on the plans. Detail a stud layout in the exact location on the top of the plate and down the sides of both plates. Leave 14½ in. between the marks, put an X where the studs will fall and write MC on both plates (see the drawing on p. 93). When the plans show the MC in the corner, mark the first stud 3 in. (two studs) out from the corner, measure over 14½ in. and mark the second stud. This will leave room for the door of the MC to open freely. When the plans show that a larger MC is to be used, check its size in the plan specifications or scale it.

Bathtubs and tubs with showers also usually come in a standard size (30 in. wide by 60 in. long), but check with the supplier. A fiberglass tub/shower requires an extra flat stud for vertical backing at 30 in.

Here the plates have been detailed so that a tub access will be framed in, allowing the plumber room to connect the tub to the drain.

so the outside flange can be nailed to it. Measure out from the inside corner 32 in. and make a mark on both plates with keel. The X for this stud location will fall away from this mark. Then back toward the inside, mark the location of a flat stud that will nail right against the first stud. The same detailing will often be needed when a tub has tile over it.

A 12-in. by 12-in. access hole through the wall may need to be provided for the trap on a bathtub, especially when building on a slab, to allow plumbers to hook up or repair the tub drain. Check the plans or ask the plumber where the drain will be. At this location, measure out 15 in. from the inside wall to the center of the tub and make a mark on the plates. Measure over 6 in. from this center mark and make a stud layout on both plates, leaving 12 in. in the clear. Later a small wood or metal frame with a door will be installed between these two studs to allow access to the tub trap.

When wall heaters are going to be installed, their location is usually noted on the floor plan. These heaters most often fit into the 14½-in. space between studs. Detail their location on the two wall plates so that the studs are properly placed and the framer won't cut in a wall brace through this area. Eventually, after the roof is sheathed, the top plate and double plate will be cut out so that the heater vent can extend up through the roof.

Cathedral or high ceilings may have beams, supported by posts or extra studs, to carry the rafters. Post location is usually found on the floor plan and post length on the sections or elevations. Both location and length need to be marked on the plates. Also, note the length of studs needed to frame any wall that is higher or lower than standard.

The plans may not show a post under all beams, especially those that frame at ceiling height, even though the posts are required by code. The important point to note is that every last framing detail will not be shown on the plans. Knowing how to handle these details comes from using common sense and experience, studying the code and communicating with other carpenters, building inspectors and other tradespeople.

Marking stud location

Some older carpenters can still remember marking stud location with a 6-ft. folding rule. Rollout metal tapes helped to speed up this repetitious process considerably, and over the years a number of other detailing tools have been devised. One ingenious tool is a circular drum that inks a stud layout every 16 in. The favored tool today, however, is the layout stick. For many years this device was made on the job (see the sidebar below), but it is now also available commercially (Pairis Enterprises, 2151 Maple Privado, Ontario, CA 91761).

When laying out stud location with the layout stick, be consistent. Keep all marks on the same side of the plates. Exterior walls should be detailed on the outside along with the marks noting door and window location; interior walls are detailed on the same side as those showing door location.

Stud detailing on walls is fairly simple. Production framers follow a couple of basic guidelines that differ from the methods of traditional carpenters. First, use as few studs as possible while complying with the building code and without weakening the building structurally. Often studs that are marked at 16 in. o.c. throughout the building are put in where

SITE-BUILT TOOLS

Making a layout stick

For the simplest layout stick take a piece of ¾-in. by 1½-in. plywood that is 49½ in. long. Attach flat strips of plywood (1½ in. wide and 9½ in. long) to the plywood stick every 16 in. o.c., with legs sticking down 3 in. on one side and 5 in. on the other side. The 3-in. legs allow you to detail the top and bottom plates at the same time. The 5-in. legs make it easy to mark two plates laid side by side or a bottom plate with a top plate set on edge where there are pipes in the wall. An extra leg can be attached at 24 in. o.c. to allow you to detail stud location at this distance.

A common, and easily understandable, mistake is to make the stick 48 in. long instead of 49½ in. Entire buildings have then been laid out slightly less than 16 in. o.c. Even when such a mistake is noticed before framing starts, it's probably best not to correct it. A few studs would be saved, but the added layout marks would be very confusing to the framers. Not every mistake needs to be corrected.

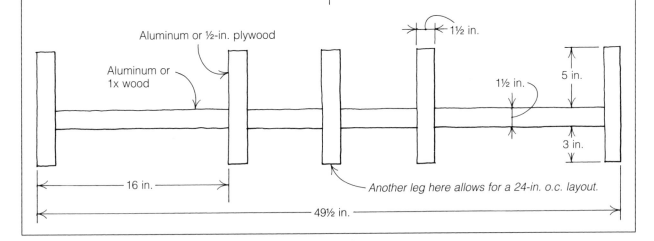

Detailing Walls for Stud Location

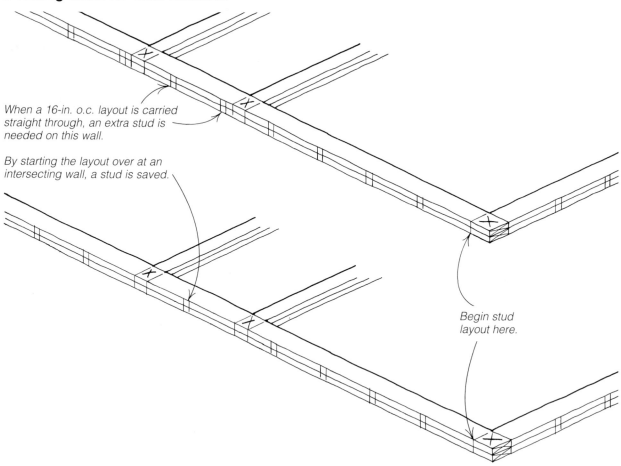

When a 16-in. o.c. layout is carried straight through, an extra stud is needed on this wall.

By starting the layout over at an intersecting wall, a stud is saved.

Begin stud layout here.

they are not needed. Fewer studs means lower lumber costs and shorter installation time.

Second, the stud location can be started anywhere and can be changed constantly. It's not necessary structurally to start marking studs at one corner and carry this detailing 16 in. o.c. along an entire wall. It might be argued that a consistent on-center layout facilitates sheathing and drywall installation. But if an expansion gap needs to be left between the sheathing, then the sheets won't always break over these studs anyway. And since most drywall installers frequently use 10-ft. or 12-ft. sheets that can span an entire room, they aren't too concerned about a regular stud layout. Finally, when you install insulation batts, fewer studs means fewer stud cavities to fill and less insulation to cut.

With these guidelines in mind and with layout stick and pencil in hand, begin the stud detailing at any outside corner. Start from the inside of the corner, not from the outside. Set the stick on the plate so that the legs go down over the top and bottom plates. Scribe along both sides of each leg at the 16-in., 32-in. and 48-in. points, move the stick forward and repeat the process. There is never any need to mark the first stud at corners and channels, because the way they are constructed there will always be a stud at this location.

Before long you will come to a door, window or intersecting wall. Every time you hit an interruption like this, start the stud detailing anew. At a door or window, start detailing at the end of the header and mark not only the two plates, but also the header and the window rough sill to indicate where to nail the cripples. At the king stud on the far side of the

Starting from the inside of the corner, begin marking the plates for the location of wall studs using a layout stick.

opening, start over again with the layout. The first stud will be 16 in. away from the king stud. Work in a similar way when you encounter an intersecting wall.

Codes in some parts of the country call for a stud under a break in the top plate. If this is true in your area, start the layout over at this point. By changing the layout constantly an appreciable number of studs can be saved, even in a small house. In larger buildings, hundreds of studs can be saved this way.

It is a good idea to detail stud location away from corners and channels. For example, if a stud on regular layout falls within 4 in. of a channel, move this stud a full 16 in. away. This will leave the area around the channel open. This is a minor detail but an important one. During framing, nails have to be driven though these corner studs, tying them together with the intersecting partition. It simply makes this nailing easier when wall studs are not in the way. And try to center any pipes between two stud marks.

There's no need to put a stud hard up against a pipe. Leave the plumbers room to finish their rough work.

Detail all the walls, headers and rough sills. When you have finished, visually check that every room has a door and a window, and that all the corners and channels are marked. The time you take to inspect the work now can save a lot of time later. It's a slow and aggravating process to have to tear out studs to put in a door that was missed while detailing. And it's fine to take a little time to admire all the work you have done so far.

BUILDING AND RAISING WALLS

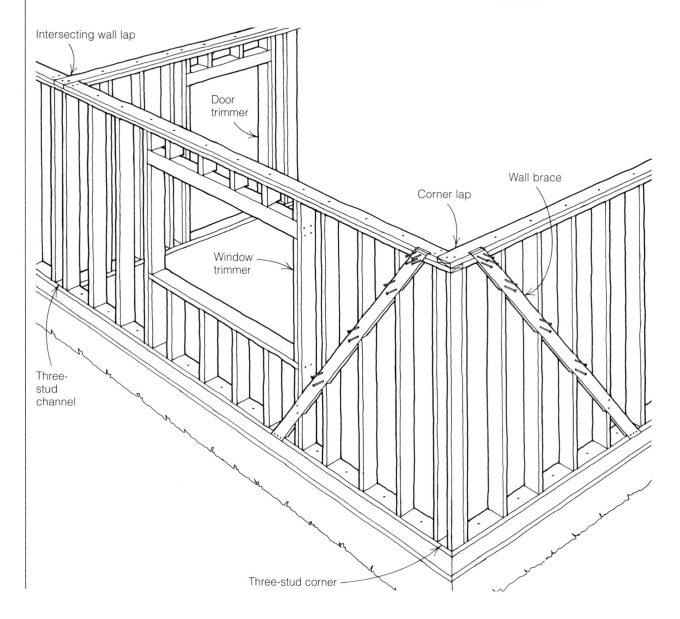

Intersecting wall lap

Door trimmer

Window trimmer

Corner lap

Wall brace

Three-stud channel

Three-stud corner

O nce all the plates have been detailed, it's time to pick up your hammer, grab some nails and start connecting the lumber to layout marks. It's time to build walls. Working on the floor, you will nail the headers, cripples, studs, trimmers, corners and braces between the top and bottom plates. The walls will be raised one by one to an upright position to form the frame of the house, which will then support ceiling joists and roof rafters or a second floor.

Older carpenters can remember how they used to nail the corner posts to the bottom plate, string the top plate across and then work off a ladder to nail studs to it. But it is much easier and faster to build walls flat on the floor. You may have to build walls on top of other plates, but often you can pull the plates of interior partitions out of the way temporarily so that the wall you are working on can be laid out flat on the floor without any obstruction.

Begin by cleaning up the work area. Except for some short 2x stock that can be used to block up plates when building over interior walls, no extra material is needed on the deck. It is especially important not to leave lumber scraps lying around with nails sticking up in them. Take a moment to bend the nails over or pull them out.

Cost-conscious builders may want to know whether it is more efficient to nail the frame together with hammers or with pneumatic nailers. The answer depends on the way job production is organized and who does the work. Some studies on cost control indicate that a good carpenter can build walls with a hammer as well as, or better than, with a nailer. Part of the problem in framing with nailers is that the hoses tend to get wrapped around different objects and have to be unraveled from time to time. Further, nails for nailers cost more than regular 16d sinkers.

Lots of studs are needed to frame a building. They are delivered to the job site in large bundles called lifts (see the photo on p. 22), which are usually held together by metal bands. Take care when cutting these bands as they can snap back and put a bad slice in an unsuspecting part of your body. Make sure that the lumber company drops the studs close by the foundation. It wastes time and energy for framers to have to carry lumber a long way.

Windows and doors

As a rule, every door header needs top cripples and every window header needs both top and bottom cripples. These have already been cut (see p. 86), so grab an armload and carry them to their proper locations. A 4⁰ header, for example, takes five cripples, so place the correct number of bottom cripples against the plates where the rough sill goes and the same number of top cripples up where the header will go during framing. Lay the cripples perpendicular to the plates. This is important. You want to place them so that they can be grabbed and nailed on in one easy motion.

Also scatter continuous window trimmers if they are being used, and window king studs, one on each side of the header. Don't worry about floating window trimmers at this point; they will be cut later.

Scatter all top and bottom cripples to their respective door and window headers before beginning to nail them on.

Nail all the cripples to the headers and rough sills before building the rest of the walls.

It is most efficient to nail together all the pieces making up the window frames first before framing the wall.

Once the window pieces have been scattered, move the header up into position, set it on edge and remove the rough sill. Now grab a top cripple, set it on the header, back it up with your foot and start two toenails. Then place the cripple on the end of the header and drive the nails home. Codes usually call for top cripples to be toenailed on opposing sides with three 16ds or four 8ds. Grab another cripple, hold it against the first, start the nails, position it and drive the nails home. Repeat the process until one cripple has been nailed on each end and to each lay-out mark on the header. A practiced framer will put two nails on one side of all cripples and then turn and nail one or two more on the other side.

Next, nail the rough window sill to the bottom cripples with two 16d nails. This is easiest if the cripples are resting against the plates. Keep the nails wide apart on the rough sill so that they will hold the cripples more securely.

If a continuous trimmer is being used, hold the bottom of it flush with the end of the cripples and drive two 16ds through the trimmer into the end of the rough sill. Now bring up the king stud, and

Windows and Doors

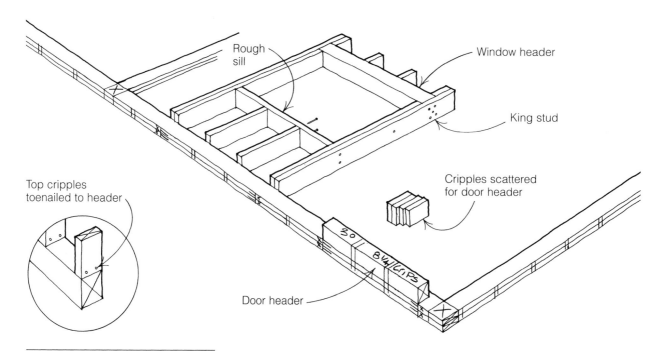

Rough sill

Window header

King stud

Top cripples toenailed to header

Cripples scattered for door header

Door header

drive one 16d nail through the trimmer into the king stud just below the header. This will do for now. More nails will be needed when the actual window frame is set. Some framers nail the trimmer or the king stud to the bottom cripple, but this is unnecessary and these nails often interfere with a saw cut later when installing a let-in wall brace.

Next, hold the king stud flush with the top of the top cripples and nail it into the end of the header. This usually calls for four 16ds for a 4x6 header and up to six for a 4x12 header. Leave this frame in position and move on to the next header.

Nail all of the members together tightly. Gaps left around the window frame can mean cracks in the drywall later on as the lumber shrinks and the building settles into its role of being a home. Nothing, not even time, is saved by being a sloppy framer. Speed, and craft, has a lot more to do with skill than it does with carelessness. Do it right and you'll only have to do it once.

Building walls

Once all the door and window frames have been assembled, it's time to put the walls together. Begin by scattering studs. If you ordered precut studs (see p. 23), they won't require any cutting to length. The order in which the walls get nailed together depends on how the building was plated. Most framers prefer to begin with the long outside through walls.

On a large building you can scatter studs to several walls. The rule is to do as many walls as possible at one time without creating a mess. Cull out any studs that are badly bowed or twisted or have large knots in them. These can be cut later for blocks or backing. Place the studs on the floor, perpendicular to the plates, one stud per layout and three studs for each corner and channel.

Now, using your hammer claws, pull the top plate away from the bottom plate. It's a good idea to bend over the 8d nails that were used to tack the plates together. If you don't, sooner or later when you bend down to grab the top plate to lift the wall, you are going to get a puncture wound in your hand. Wounds and bruises are part of the trade, and every framer has his or her share. Sometimes you hear carpenters say that they can't afford to donate blood to the Red Cross; they have none to spare because they

Once the studs have been scattered, you can begin nailing the wall together. Pull the top plate into position, line up the first stud with the layout mark, then drive two nails through the plate into the stud.

lost too much on the job. That may be a bit of an exaggeration, but does point out that everyone needs to be safety conscious.

Beginning at the outside corner, nail the first stud to the top plate with two 16d nails, flush with the end. Hold the nails apart, one high and one low, just as you nailed the rough sill to the cripples. Then, on a 2x4 wall, nail two more studs alongside to form this corner (see drawing A on p. 102). These three studs will be nailed to each other when the bottom plate is nailed on. At channels a stud is nailed in flat between two regular studs to give backing to an interior partition (drawing D).

Corners and Channels

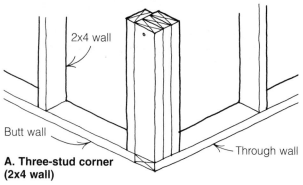

2x4 wall

Butt wall

A. Three-stud corner (2x4 wall)

Through wall

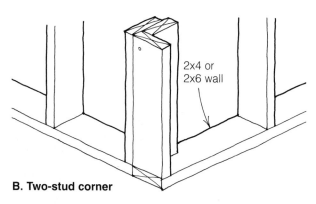

2x4 or 2x6 wall

B. Two-stud corner

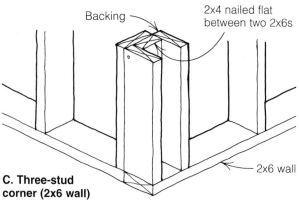

Backing

2x4 nailed flat between two 2x6s

2x6 wall

C. Three-stud corner (2x6 wall)

D. Three-stud channel

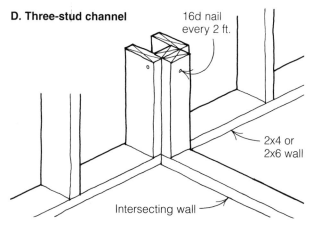

16d nail every 2 ft.

2x4 or 2x6 wall

Intersecting wall

A three-stud corner can also be built by using scrap blocks rather than a full stud in the center. But if you're building a lot of corners, you'll find that it is quicker and requires fewer nails to use a full-length stud.

Some framers like to save a stud at the exterior corners by building a two-stud corner (drawing B at left). The first stud is nailed in on edge flush with the end of the plate and the second stud flat alongside the first so that when the wall is raised it will give backing to the intersecting wall. The two-stud corner is a legitimate corner that works quite well and can be used when allowed by local codes.

Corners and channels for 2x6 walls are built in much the same way as for 2x4 walls. One difference is that exterior corners are framed with the first 2x6 nailed in on edge, then a 2x4 is nailed in flat, and another 2x6 is nailed in on edge (drawing C). This design offers a corner wide enough to give full backing to an adjoining 2x6 wall.

Once the corner studs are in place, it's just a matter of continuing to frame on down the line, matching lumber to layout marks and nailing it in place. Pay attention to what you're doing. If the layout, plating and detailing were done correctly, this should be a pretty routine job.

It can be a real pleasure to watch an experienced carpenter nail together a wall. A good framer will develop a rhythm or pattern while working. The whole body moves in a fluid, not forced, manner. It's not just the arms and hands that are working but also the feet. Trained feet can move studs into position for nailing and even lift up a stud slightly to hold it flush with the plate while it is nailed in position. Yet anyone who has ever framed walls for a full eight-hour day knows it is hard work, especially when working on a plywood deck on a hot day. Drink a lot of water so you don't develop what framers call "vapor lock" and medics call heat exhaustion.

Once all the studs, headers and cripples have been nailed to the top plate, pull up the bottom plate and nail it to the studs and cripples. When framing on a slab with foundation bolts through the sill, be sure not to turn the plate over as you nail it to the studs or the holes in the sill won't line up

Taking care of your body

Some state legislatures have outlawed the short-handled hoe for farm workers because it is too hard on backs to bend over low all day long, but no one has considered outlawing the short-handled hammer for carpenters. It is hard on backs to bend over and frame all day long. The long-handled hammer does help, but you still have to reach down to hold the nail so it can be set and driven. There is no way to frame efficiently standing upright. Part of being a good carpenter is learning how to take care of your body. Learn to lift with your legs instead of your back. Exercise to keep strong, soak in a hot bath regularly and get a good massage occasionally to loosen up tight muscles. Many lower-back injuries happen on those cold mornings when the body is still tight. Few carpenters will do calisthenics out on the job, but there is no reason why you can't do some stretching at home before leaving for work.

Nail the top of the end stud on a butt wall about ¼ in. from the edge of the top plate to ensure that the top plates of adjoining walls will fit tightly together.

with the bolts. The bottom plate will usually nail on faster than the top, because all the studs are on edge and more or less in position. Now's the time to nail the corner and channel studs together with three 16ds on each side. Drive the first one 2 ft. from the bottom and then two more about every 2 ft.

Intersecting or butt walls are built after the exterior walls have been raised. These frame just like the through walls. A minor difference is that if you nail the end stud back from the end of the top plate about ¼ in. (see the photo at right) it will allow the top plate of the butt wall to join freely with the top plate of the through wall once they are raised. The top plates of raised walls must fit together tightly — no gaps are allowed.

Rake walls

Building rake walls requires that the top ends of the studs be cut at the roof-pitch angle. During the layout stage, a full-size rake-wall layout was marked out on the floor (see pp. 69-70). Using straight stock, scatter studs that are a little overlong at every layout on the bottom plate. Then nail the studs, along with all headers, sills and cripples, trimmers and king studs, to the bottom plate just as on a regular wall. Position top cripples on top of the headers. Framers prefer not to nail the top cripples to the header until they are marked to length and cut.

Set the top plate on edge on the chalkline and mark the stud layout on it.

Next, straighten up the bottom plate and hold it in place just below the chalkline by tacking it to the floor with a few 8d nails. Then bring up the top plate, lay it across the studs and position them on the same layout as the bottom plate, ready for marking.

The next step is to snap a chalkline on the studs directly over the roof-pitch line marked on the floor. Before cutting the studs, grab a longer piece of plate stock. A new top plate has to be cut because it runs at an angle and is longer than the original plate. Lay it on edge directly on the chalkline and mark it for length. Now, with the plate still on edge across the studs, take a pencil and mark the stud location so that the studs can be nailed in once everything is cut.

If the plans call for a ridge beam at the top of a rake wall to carry the rafters, it will sit on a post, often bolted or nailed to it using a metal post cap. The length of the post was determined during layout (see p. 70), and it can be cut and installed now.

Now it's time to cut the studs to length. Set the saw to the correct degree for the roof pitch (for example, 18½° for a 4-in-12 pitch), pick up the studs one by one and make the cuts. Saw tables tip only one way. If the cut angle runs in the opposite direction from the tilt of your saw, cut the stud to length square and then tilt the whole saw over and make the angle cut. Finally, nail the cripples to the header, then nail the top plate on the cripples and wall studs.

Once the rake-wall studs and cripples have been marked to length, they can be cut in place.

Double top plate

The next step is to nail the double top plate to the top plate. The double top plate ties all the walls together and makes the entire building structurally stronger. Traditionally, carpenters would nail on the double plate after the walls were raised, working off ladders. But this method is very inefficient; double top plates can be nailed on much more quickly and easily while the wall is still lying on the floor.

Begin by scattering the plates right on top of the wall, near the top plates. This way no measuring tape will be needed to cut the plates to length. Try to use long (16-ft. to 20-ft.) and reasonably straight stock. On through walls, hold the double top plate back from the ends the width of the framing stock (e.g., 3½ in. on 2x4 walls). Do the same at all channels. When the doubles are cut for the intersecting walls, they will be cut long and then lap over these gaps to help tie the walls together.

Where walls of different heights intersect, there is no need to leave cutouts in the double top plate. They will have to be connected later by metal plate straps (see p. 115). On rake walls, the double top plate hangs over on the low end about 3 in. so that it can lap and tie into the wall it butts. At the top end, let it run long and cut it when the ridge beam is set on the post.

Most building codes require that double top plates lap over joints in the top plate by at least 4 ft. There are times when this isn't possible, either because the plate stock isn't long enough or the intersecting walls are too close together. In such cases, nail an 18-in. metal plate strap with four 16d nail holes in each end over the joint in the top plate.

Double Top Plates

Long walls can be broken into sections for easier raising by leaving the double top plate lap unnailed.

Corner gap — Nail over studs. Channel gap —

When ducts for heating or air-conditioning systems need to pass through the wall, you will need to cut out the plates for them. The location of these passages was marked on the top plates when they were detailed. End the double top plate on each side of the required opening; leave the top plate intact for now and cut it after the walls have been joisted.

Start nailing on the double top plates by driving two 16d nails at the end of each plate and then one more over every stud. If you drive nails between studs you run the risk of hitting one with a sawblade when cutting in braces. Also, electricians and plumbers like to run their work through these cavities. Anything that can be done to help other tradespeople will make the entire job run more efficiently. Make sure that the ends of the top plates butt together tightly. A gap between top plates will lengthen the wall and make it impossible to plumb at both ends.

Cutting trimmers

Floating window trimmers and trimmers for prehung doors can now be cut and nailed in. When conventional jambs are used, door trimmers are often installed later when setting door jambs to ensure that the trimmers will be straight. Slight bows are tolerable in most framing, but straight stock is needed for trimmers because they support the finish door jamb. The difference between prehung and conventional jamb trimmers is that the first can be nailed and held securely in place during wall framing whereas the latter are secured when the jamb is set. If the plans require that exterior walls be sheathed (see pp. 125-126), both door and window trimmers need to be cut and set exactly in place before nailing on the sheathing panels.

Trimmers need to fit snugly under the header, but this cut too can be gauged by eye. Hold the 2x trimmer stock against the header, rough sill or bottom plate, sight it to length at the underside and cut. The sawblade should graze the edge of the header or rough sill for a perfect fit. With a little practice, you should be able to make these cuts perfectly. If it makes you feel more comfortable, however, you can mark the cut with a pencil to help ensure a tight fit.

Blocking and trimmers can be cut without measuring and marking. Place the stock squarely in one side of the opening and eyeball the cut along the other side.

Insert the window trimmer under the header and drive one 16d through its center into the king stud; additional nails will be driven when the window is set. Nail the prehung door trimmer through the bottom plate and then about every 16 in. through the trimmer face into the king stud; once drywall is installed, the door frame can be set.

Special blocking

Additional blocking in the walls may be required by code or the plans for a variety of situations. Fireblocking, for example, is usually required in walls over 10 ft. high. When fireblocking is needed, cut the blocks to length and nail them in at staggered heights for easier nailing. It's best to nail them either high or low in the wall, because blocks nailed in at head height are likely to cause injury to someone passing from room to room through the studs. Back when fireblocking was the rule, older carpenters can recall having their eyes crossed by walking into a block installed head-high.

In areas prone to earthquakes, the code or a structural engineer may specify that the drywall be installed with extra nails or screws to increase the stability of the wall. This may require extra blocking. For example, the plans may call for a row of blocks to be run in 4 ft. high on an 8-ft. wall. In this case, snap a chalkline along the wall 48½ in. from the top of the double top plate. Then, with the wall still flat on the floor, nail the blocks in, centered on the chalkline. The drywall will then be installed horizontally, with plenty of backing for the extra nails or screws.

Walls that are going to be covered with vertical siding without sheathing underneath may need rows of blocks run horizontally for backing. Snap a chalkline every 2 ft. and nail in the additional blocks.

Two blocks are needed in walls to provide drywall backing at the top and bottom of a medicine cabinet. The medicine cabinet was located between studs by marking "MC" on the plates during detailing. Hook your tape over the bottom of the bottom plate and mark the vertical studs at 4 ft. and 6 ft. Nail blocks at these marks, leaving a 14½-in. by 24-in. hole into which the standard medicine cabinet will fit.

Backing for smaller bath and kitchen items such as towel racks, paper holders and grab bars is usually left until later. It is often easier for the builder to locate these items when the room has been completely framed.

Wall braces

If the walls aren't going to be sheathed or sided with plywood, they will need to be braced laterally with let-in braces made of metal or 1x lumber. Wall bracing helps to hold the building plumb (straight up and down). When bracing is required, building codes usually call for permanent braces in almost every wall that has room for one. On longer walls, braces are required at each end, with an additional brace for every 25 ft. of wall space. If there are two braces in a wall they should oppose each other, one slanting one way, the other slanting the opposite way.

Even though permanent bracing is not necessary on a wall that will be sheathed, temporary braces will be needed to plumb and line the walls (see pp. 116-124). Because temporary braces often get in the way, many builders go ahead and install permanent ones even when they are not required by code.

Metal angle framing braces are a fairly new item in the construction industry. Compared to 1x6 let-in braces, metal braces are faster, easier and safer to install and will hold the building plumb just as well. Flat metal braces can also be used to provide lateral stability to walls. But, unlike the angle braces and let-in wood braces, they must be nailed on in pairs, forming an X to stabilize the wall in both directions.

To install a metal angle brace, lay it across the framed wall diagonally from the bottom plate to the double top plate and make a pencil mark along one side. Then cut a slot 1 in. deep along this line into the plates and studs. Slip one flange of the brace into the slot and nail it to the bottom plate with two or three 8d nails, then add one more 8d through the brace and into the first stud. At the double top plate, start an 8d nail alongside the brace and bend it over to hold it in place as the wall is raised. Later, when the building is plumbed, it will be nailed permanently to the other studs and plates.

If metal braces are not available you can let in a 12-ft. 1x6 or 1x4, as the code allows. An experienced framer can cut in one of these in a matter of two or three minutes. They are good braces, but the procedure for installing them needs to be studied well and applied carefully to avoid the possibility of injury.

Make sure that the wall is reasonably square. Then lay the brace at about a 45° angle across the studs, avoiding door and window openings. Set your sawblade to a depth of about 1¾ in. Place one foot on the brace to keep it in position and hold the saw

Permanent Wall Braces

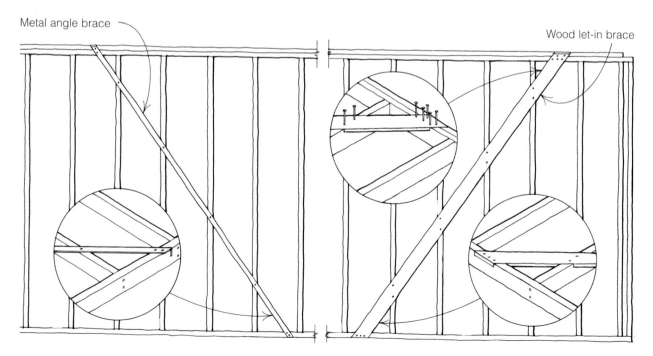

Metal angle brace

Wood let-in brace

with both hands to avoid kickback. First, cut the end of the brace flush with the bottom plate. If you are working on a concrete slab, trim the bottom end about ½ in. short to keep it from contacting the slab directly. Next, with your saw riding on top of the brace, cut a slot 1 in. deep on both sides along the studs and plates. Cut the brace ½ in. short of the top of the double plate.

Now the wood between the two cuts must be removed. One way to do this is to drop the sawblade down into the plates and studs several times to a depth of about 1 in. The wood can then be removed by chiseling at it with your straight-claw hammer. But this method is slow and leaves an unclean cut.

The better and more efficient way to notch the studs requires care and experience to do it correctly and safely. First, lay the saw over on its side, place the front edge of the saw table on the stud and use the thumb of the left hand to operate the trigger. Until you gain experience at this technique, brace your left elbow against your knee so that you will be able to stop the saw from kicking back while cutting in this position. With the right hand, grab the top handle of the saw after lifting the guard slightly. The object is to drop the blade into the stud or plate and cut out the slot so that the brace will fit. Just remember that the slot needs to be at least ¾ in. deep. Here, as elsewhere, a sharp sawblade makes for the safest cut.

To cut out studs for a 1x let-in brace, lay the brace stock along the wall in its intended location. Using the brace as a guide for your saw, cut along both sides.

For the final cut, place your saw on edge and drop the blade into the stud. Be sure to keep both hands on the saw.

Now drop the 1x brace into the slot and nail three 8ds through it into the bottom plate, then two more into the first stud. At this point, nail only at the bottom. Start two nails in the brace at the remaining studs and five more at the top and double top plates. These nails will be driven home when the walls are raised and plumbed.

Nail some 2x scraps to the rim joist to keep the wall from sliding over the edge when it's being raised.

Rake walls and other tall walls can be braced by using a longer 1x or two metal angle braces. If you're using the latter, you will need to overlap the braces at the center, which will require cutting a slightly wider slot in the studs.

Raising walls

Once you have nailed all the long through walls together, you can begin to stand them upright. Most wall raising is still done by simple human labor, although some pretty good mechanical devices are available to help with the process. The weight of any given section of wall depends on its size and the wetness of the lumber. An average-sized person can usually raise a 10-ft. section of a standard 8-ft. high, 2x4 wall.

If not enough bodies are available at raising time, you may need to cut the wall into sections before it can be raised. With one end of the double top plate loose, cut the bottom plate so that it breaks directly under a break in the top plate. Once the wall is raised, the lap of the double top plate can be nailed to tie these two sections together.

Begin by cleaning up any debris on the floor where the bottom plate will sit. Walls are held upright by temporary braces, one on each end and another about every 10 ft. between. Use studs for braces and scatter them down the wall before raising. The end braces can usually be nailed to a corner stud before raising; the remaining braces will be nailed in place once the wall is upright.

To keep outside walls from slipping over the edge while being raised, nail some pieces of 2x stock to the rim joist so that they stick up above the floor a few inches to catch and hold the bottom plate (see the photo at left). If you are working on a slab, the foundation bolts will serve this purpose.

Raise walls using your legs and arms, not your back.

It's easier to get a secure grip on the wall if it's not lying flat on the deck. So stick your hammer claws into the double top plate, pick it up a bit and kick a 2x block under. Some framers lean a short 2x block against the wall at about a 45° angle, sink hammer claws into the double plate, then raise the wall enough to allow the block to fall under the plate. Keeping your back straight, lift the wall to your waist using your legs, then overhead with your arms and upper body. Then, by pushing on the studs, continue to raise the wall until it's fully upright. The wall can easily be held in this position until temporary braces are nailed on. Before nailing the bottom of the braces, a smart framer will lean the main exterior walls out a little bit, making it easier to raise the interior walls. Two or more persons will need to steady a long wall while others nail on additional braces. Take the extra studs that were scattered for braces and nail them to a wall stud about every 10 ft. and down from the top plate about 2 ft. Place the brace stud flat against the wall stud and nail the two together with two 16ds, then nail the lower end of the temporary brace through the subfloor, into a joist below, with two more 16ds.

Hold the wall upright until temporary braces can be nailed in place.

Temporary Wall Braces

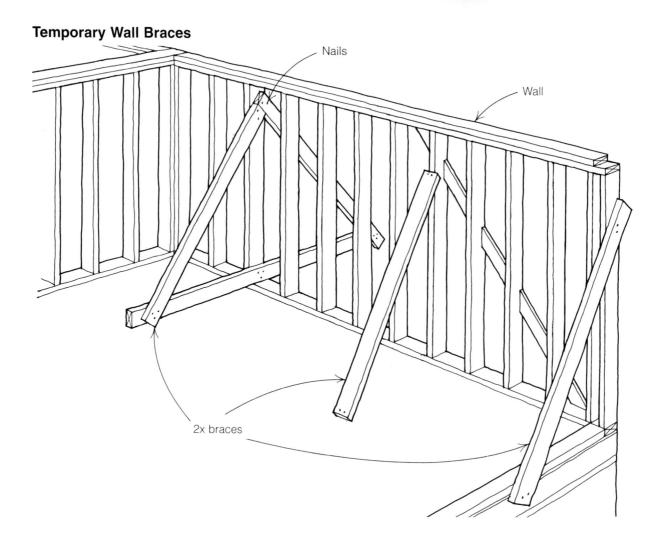

Nails

Wall

2x braces

If the walls are going to be left like this overnight or longer, without intersecting walls being tied into them, take the time now to put on some sturdier braces. Always prepare for the worst: Poorly braced walls have been known to get blown over the side by a sudden wind gust. One simple way to brace the walls is by shoving a long 2x halfway through the stud wall, nailing it on edge to a stud at the bottom plate and then forming a triangle by nailing 2x4s from each end of the long 2x up to the top of the wall (see the drawing above). Blocks can also be nailed to a wood floor and braces nailed from them to the wall. If you are working on a slab, attach some 2x4 blocks to the floor with pins driven by a powder-actuated fastener (see p. 114) and secure braces from these to the top of the wall.

Make sure that the raised wall aligns properly with the layout line. You can move a wall by sinking the claws of your hammer under the bottom plate and prying it. A sledgehammer can also be used to tap a wall into position. Finally, nail the bottom plate to the floor with one 16d nail per stud space. Drive a nail right next to every king stud, but don't nail in doorways because the plate will be cut out later when the door jamb is set.

When raising walls on a concrete slab, first shove the bottom plate against the foundation bolts. Try to align the holes in the plate with the bolts so the raised wall falls into position. If, despite your efforts, the wall lands on top of the bolts, it will need to be moved. This can be done easily with a short piece of 2x stock. Put the 2x under the plate, use it as a lever to pick the wall up and move it into position

and drop it down over the bolts. On longer walls this operation may take two or three people working together.

Once the through walls have been raised, the shorter butt walls that form other outside and partition walls can be built and raised. Intersecting walls often have to be raised one end at a time. Pick up one end enough to clear the through wall and then raise the other end. You might find it easier to fit the lapping double top plate on the intersecting wall into the gap in the through wall if you tap it loose from the top plate about ½ in. before raising. Remove the end temporary brace holding the through wall, make sure that the bottom plates of the two walls are both flat on the floor and then nail the end stud of the intersecting wall to the corner of the through wall. Evenly space three 16d nails up the end stud and into the backing studs. If you are working on a wooden floor, check that the wall is on the layout line and then nail it down. Walls raised on concrete will be secured to the line later. Continue building and raising the rest of the walls.

Walls raised on a concrete slab may need to be moved to align the holes in the plate with the anchor bolts.

Intersecting walls are raised and tied into other walls.

Exterior walls on a concrete slab are secured with nuts tightened on the bolts.

Securing walls to concrete floors

On a slab foundation, the exterior walls are usually bolted to the floor. Washers are placed over the bolts and nuts are then tightened with an impact wrench, a socket wrench or a simple adjustable wrench. Before tightening the nuts, make sure that walls sit directly on the chalkline so that they will be straight and easily plumbed. If they don't, use a small sledgehammer to line them up.

To fasten interior walls to the floor most carpenters now use a powder-actuated fastener, which fires a pin through a metal washer, through the bottom plate and into the concrete. This tool should not be used on exterior walls, because the force of the pin can blow a hunk of concrete off the edge of the slab. Before operating a powder-actuated fastener on the job site you should get training from a qualified instructor and a license. This is a tool that demands respect. Read the directions carefully and always wear proper eye and ear protection; never point the fastener at anyone and use it only in the way it was designed to be used.

Interior walls on a slab foundation are often secured with a pin driven by a powder-actuated fastener.

Be sure you have the right-size pin. Codes often call for a 3-in. long pin placed 1 ft. from each end of a plate and then every 6 ft. for the rest of the length of the wall. Don't put any pins in doorways, but do put one on each side of each door, close to the king stud, which will hold the door frame secure once the door is hung.

Some regions allow interior walls to be secured to the floor with a special hardened nail that is driven in with a hammer. Check your local code.

Corner and channel laps are nailed to intersecting walls with two 16d nails through the double top plate. Make sure that all walls butt tightly and are properly aligned with their layout marks.

Tying off double top plates

It doesn't take long to nail the double top plate laps at corners and intersecting walls on an average-sized house. But it is important to do it right, because it will make the walls much easier to plumb and line. The fastest way to do this job is simply to hop right up on the walls. It always comes as a welcome relief from framing to jump up on the walls and tie off the laps.

The most important parts of this job are making sure that the plates are nailed in the correct position and that the top plate of intersecting walls butts tight to the top plate of through walls. All the channel and corner marks that were put on during detailing are now used as guides to ensure that everything comes together properly. If necessary, a 16d toenail can be driven through the edge of the double top plate of the butt wall and into the top plate of the through wall to force it to line up with the corner or channel marks.

Now you'll appreciate having held the last stud ¼ in. away from the end of the top plate on butt or intersecting walls (see p. 103). The gap allows the top plates to come together freely. As soon as all joints fit tightly and top plates line up with corner or channel marks, drive two 16d nails through the lapped double top plate and into the top plate of the through wall.

Any gaps between the top plates of the butt wall and the through wall have to be closed if the building is to be plumbed properly. Usually the easiest way to close these gaps is to drive a 16d toenail into the top plate of the through wall up into the double plate of the butting wall. Striking this nail a few times should draw everything together. Now walk the plates and repeat the same procedure on all of the laps at every corner and channel.

When necessary, high walls can be tied to low walls using metal plate straps. The strap can be run across one of the top plates and secured to a block nailed between studs in the high wall.

PLUMBING AND LINING

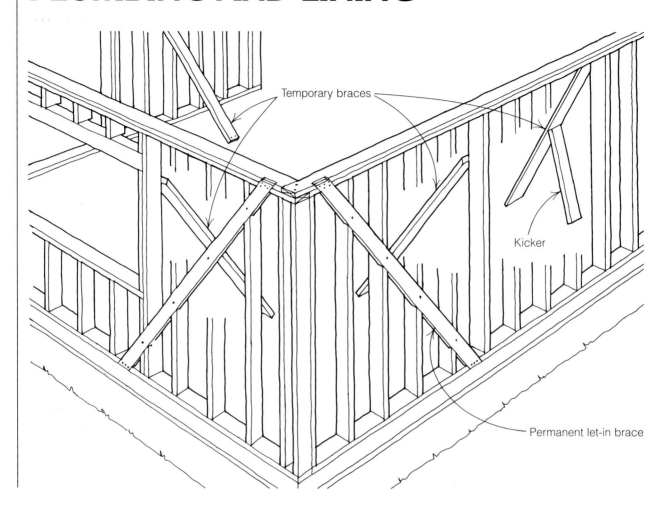

Temporary braces

Kicker

Permanent let-in brace

When all of the walls have been secured to the deck and tied together, they need to be made straight, parallel and true. Carpenters call this process "plumbing and lining" a building. This is not a difficult job, but it is a very important one. Walls that aren't plumbed and lined will result in crooked hallways, bowed walls, and doors and windows that won't fit. Cabinets are difficult to hang on crooked walls. And roof rafters will only fit properly on walls that are straight and parallel. Good framers know that if they plumb and line a building with care, everything that follows will go easier and the finished product will look better.

Plumbing simply means straightening from the bottom of the wall to the top, in other words, making the top plates line up directly over the bottom plates. Lining means straightening a wall from one end to the other. Both jobs require two people.

Plumbing is done first. But before you start, you might want to let gravity do some of the work for you. Framed walls left standing overnight have a tendency to straighten up and can be near plumb the next morning. If the wall plates were cut accurately and the framing was done properly, the walls should need little adjustment once they have settled into their upright position.

Making a plumb stick

All you need to make a plumb stick is a 2-ft. level, a 2x4 stud, some duct tape or heavy-duty rubber bands and a shim or two. Pick out a fairly straight, lightweight stud. Then cut two 16-in. pieces of 1x2 and nail them to the edge of the stud at each end, leaving 3 in. or 4 in. overhang. This way the plumb stick will rest only against the top and bottom plates and not against a potentially bowed stud, which would produce an inaccurate reading.

Attach the level to the opposite edge of the stud using duct tape or rubber bands, high enough so that the bubble in the vial will be at eye level when the plumb stick is held in an upright position.

To check the plumb stick for accuracy, hold it upright with the face of the stud flat against the wall and the 1x extensions touching the bottom and top plates. Move the top of the stick until the bubble is centered exactly in the vial. Then mark along the edge of the 1x extensions on both the bottom and top plates. Now turn the plumb stick over so that the opposite face of the stud is flat against the wall and line the extensions up with the marks on the plates. If the bubble is in the exact center of the vial, the plumb stick is accurate.

If the bubble is not centered in the vial, the level needs to be adjusted. Simply stick a wood shim, folded paper or even an 8d nail under one end of the level. Then check the plumb stick again. Keep adjusting the shims until the bubble is centered both ways.

A plumb stick can be made using an old 2-ft. level taped to a long 2x4 (usually a wall stud), with 1x scraps at the top and bottom to hold the stick away from any bowed studs.

Walls are checked for plumb with a level. You can use a 4-ft. or 8-ft. level or buy an expensive one with digital readout, but many experienced framers prefer to make a plumb stick using a 2-ft. level and a stud, as described in the sidebar above. Levels are sensitive tools, and in the rough and tumble world of framing they can get knocked out of shape and start to look somewhat like an old framer. Even a brand new level won't necessarily be completely accurate. A plumb stick allows you to compensate for any inaccuracies in a level.

To be plumb, the top of this wall must be moved to the left. Brace a bowed 1x push stick with your foot. Then, by pulling up on the bow, the force will move the wall.

Plumbing walls

Before you begin plumbing, make sure that the bottom plate is nailed directly on the chalkline. If it isn't, the wall can't be plumbed accurately. One person is needed to hold and read the level. Another person pushes on an adjoining wall until the first wall is plumb and then nails off the braces. Begin in an exterior corner and work your way around the building. The plumb stick operator tells, or signals, which way and how much to move the wall.

If the wall is already plumb, go ahead and nail off the braces. If the wall needs to be moved, you can try to give it a push. Sheer muscle will move some walls, but others will require the use of a simple push stick (see the sidebar on the facing page). Place the stick under the top plate against a stud and running diagonally down to the floor. It needs to be as close to parallel with the wall as possible so that it

will push the wall in the right direction, Bend the stick down, holding the bottom end against the floor with one foot. Now pull the middle of the stick up. As the board straightens, the wall will move. It is a simple method, but quite effective. You may not be able to move a mountain with this stick, but you can move a substantial section of wall.

If your 1x push stick isn't up to the task on heavier walls, try using a long 2x4. For extra leverage, place the upper end against a stud at the top plate and the lower end against the bottom plate of an intersecting wall. Bow the 2x down and move the lower end along the bottom plate toward the wall that is being moved. By pulling up on the bowed 2x4, you can exert enough force to move almost any wall.

The object is to rack the wall (move it laterally) just enough to make the adjoining wall plumb.

Once the wall is plumb, bracing is nailed on to keep it that way.

While one person maneuvers the push stick, the other keeps an eye on the level. As soon as the bubble is centered, the wall is plumb and the permanent braces need to be nailed on so that it stays that way.

Nailing off permanent braces is a simple process. If you can't reach to the upper ends of the top plates, work from a stool or small workbench. With metal braces, nails need to be driven through holes in the metal flange and into each stud. If the brace is on an exterior wall, you may have to lean out and wrap one leg around a stud to stabilize yourself before driving nails. Particularly on an upper floor, be sure that the stud you hold onto is properly nailed

When necessary, nails can be started with one hand. Just hold a nail against the hammerhead and strike it into the wood.

Temporary Braces

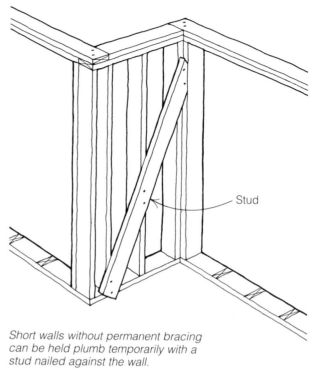

Short walls without permanent bracing can be held plumb temporarily with a stud nailed against the wall.

and free of large knots. The process is much the same for wooden let-in braces except that the nails in these braces were started earlier. All you have to do now is drive them home.

If you are leaning out on an exterior wall, holding on with one hand, and find that you need to start another nail, here's a tip on how to do it one-handed. Put your hand around the hammerhead, place a nail between the index and middle finger, and hold the nailhead against the side of the hammer. Now hit the nail onto the brace hard enough to get it started, grab the hammer by the handle and drive it the rest of the way.

Once one end of a wall is plumb it never hurts to check the far end. If the plates have been cut accurately and the walls framed properly, when one end is plumb the opposite end should be as well. If it is not, check the wall to see that all the joints in the top plate are butted tight together.

When the outside corners are done, work your way through the building plumbing every corner and every intersecting wall that you can. With the exterior walls already plumbed, most interior walls will be very close to plumb.

Permanent bracing may not be possible on short walls, or necessary on exterior walls that are going to be sheathed. In these cases, put in some temporary bracing to hold the walls plumb until the ceilings have been joisted and the walls have been sheathed. Don't hesitate to use plenty of temporary braces. Extra braces at this point will guarantee that the building is held plumb and square until the rest of the framing is completed. Temporary braces can be made by nailing the upper end of a stud in a corner and then diagonally across several wall studs to the floor. Once the wall has been pushed to a plumb position, nail the stud to the bottom plate.

Using hammer claws under a temporary brace can pry some walls into plumb position. Place the claws between the lower end of the stud and the floor and move the hammer handle down to put pressure on the stud. With a little practice, this method can move a considerable section of wall.

Bracing a Garage-Door Wall

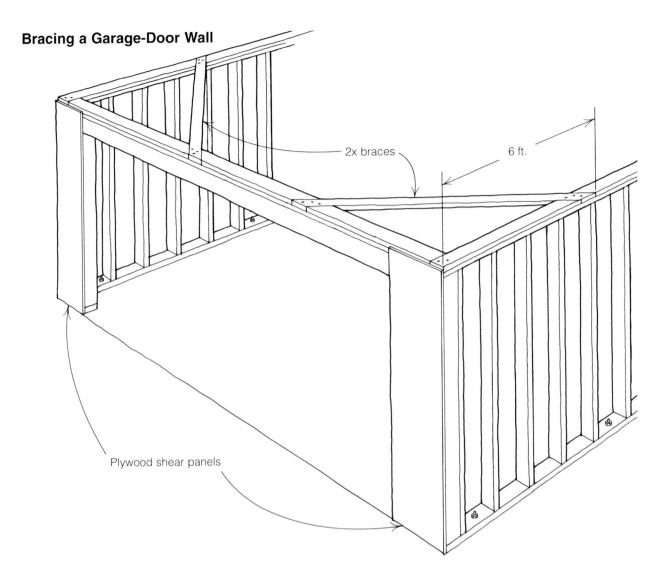

2x braces

6 ft.

Plywood shear panels

Plumbing a garage-door wall

On a typical garage, the front wall is mainly all header, allowing a large opening to park cars. Very little wall space is left on either side of the door opening; certainly not enough in which to cut a diagonal brace. There are other methods of holding walls plumb that you can use in this situation.

One way is to plumb the walls and then lay a 10-ft. or 12-ft. 2x4 diagonally across the double top plate from the wall containing the garage header to the side wall (on a detached garage, put the brace on both front corners, as shown in the drawing above). Mark the location of the 2x on the plates, cut out these sections and then drop the brace into

the slot. Nail it in place with 16d nails and trim the ends of the brace flush with the outside of the walls.

You can also nail a plywood shear panel to the small section of wall on either side of the garage-door opening. Even though these walls will generally be quite narrow, perhaps 1 ft. or so, an 8-ft. high section of plywood will help brace the entire wall.

A more traditional method of bracing garage walls is to use long or lapped 1x6s, nailed diagonally on the double top plates from corner to corner of the garage, forming a large X at ceiling height. This bracing will hold the garage-door opening stable until rafter ties are in place (see pgs. 25 and 152). The 1x braces can then be nailed up into the ties for further stabilization.

Lining walls

All walls need to be lined. Lining walls simply means straightening the top plates from one end of the wall to the other. The bottom plate is straight because it has been nailed to a chalkline. Once all the walls are plumb, the top plates should be close to straight, especially if straight stock was used. Lining can be done by eye or using a dryline, and walls can be held in place by temporary braces until the building is joisted and has a roof.

The easiest way to line a wall is to climb on top of the wall at one corner and sight down the length of the top plates. If you don't yet trust your eye, you can use the dryline method, as shown in the drawing below. Nail a scrap of 2x to the top plates at each corner and drive an 8d nail part way into the outside ends of each block. Hook a string to one of the nails, pull it tight and secure it to the nail on the other end. Now take another scrap of 2x and move it along the length of the wall, checking for a consistent 1½-in. gap between the string and the wall. Any walls that are out of line need to be adjusted and then held in place with temporary line bracing.

If the top of a wall needs to be moved out, nail a bracing stud with two 16d nails flat against the edge of a wall stud about three-fourths of the way up on the wall. If you are working on a wood floor, hook the claws of your hammer under the stud and pry until your partner says that the wall is straight. Then nail the bottom end of the brace to the floor

Checking Walls with a String

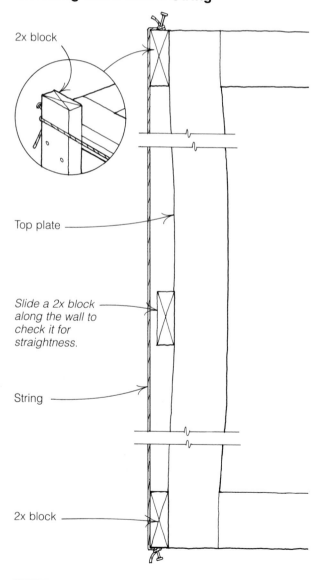

2x block

Top plate

Slide a 2x block along the wall to check it for straightness.

String

2x block

If the top of a wall needs to be pushed out a bit, nail a bracing stud against a wall stud. Pry the brace with your hammer claws until the wall is lined, then nail the brace to the floor.

with two 16d nails. Try to drive these nails through the subfloor and into a joist to make sure that the brace is fastened securely. On longer walls, nail a temporary brace in about every 10 ft. even if it sights straight.

Tall or rake walls can be braced by nailing longer 2x braces flat to a wall stud near the top and securing them at the bottom to the base of an interior wall. When walls are going to be sheathed outside, don't let braces run through the wall.

If you are working on a concrete slab, take a 4-ft. long piece of 2x, nail one end of it flat to the bottom plate of the wall with two 16d nails and let the other end extend into the room. The bottom end of the stud brace can now be nailed to this 2x, as shown in the photo below. It can also be secured to a temporary block fastened to the floor with a steel pin (see p. 114).

If the top needs to be moved in, especially on a slab, you'll have to follow a different procedure. On an exterior wall on a one-story building, nail a 2x on edge to the bottom of a wall stud. Let it extend to the outside, sitting on hard ground or a 2x block that rests on the ground. Nail a long stud near the top of the same wall stud, with the bottom end positioned on the horizontal 2x. The top of the wall can now be pushed and held straight by nailing the bottom end of the brace.

Another way to move a bowed wall in is to take a long 1x or 2x and nail one end of it flat to the edge of the top plate with several nails. Make sure that it doesn't extend above the top plate. Nail the other end to a joist through the subfloor or to the bottom plate of an interior wall. If you are working on con-

On a concrete slab, nail the bracing stud to a 2x nailer secured to the bottom plate.

At ground level, walls can be pushed in and braced by nailing a 2x to the bottom plate and running it out from the building, providing a nailing surface for the brace.

crete, nail it to the bottom plate of an interior wall. Then take a 3-ft. to 4-ft. 2x and place it upright near the middle of the brace, with one end on the floor and the other end up under the brace, as shown in the photo at left. Push the short 2x against the brace, putting a bow in it. This will cause the wall to move in. When the wall is straight, drive a nail down through the brace into the short 2x post to hold everything secure.

In hallways or narrow rooms such as closets and bathrooms, one wall can be straightened with temporary braces and a parallel wall can then be held straight with a 2x nailed near the top of the studs. Take a scrap of 2x and place it on edge between the parallel walls at the bottom plate. Mark the 2x on each end, move it up to the top plate and nail it in position following the pencil marks, which indicate exactly how far apart the walls should be.

If the top of a wall needs to be pulled in to make it straight, nail a 1x brace from the top plate to the floor. Wedging a 2x kicker under the brace, draws the top in.

Parallel walls can be held straight by tying them together with a 2x nailed at the top.

SHEATHING WALLS

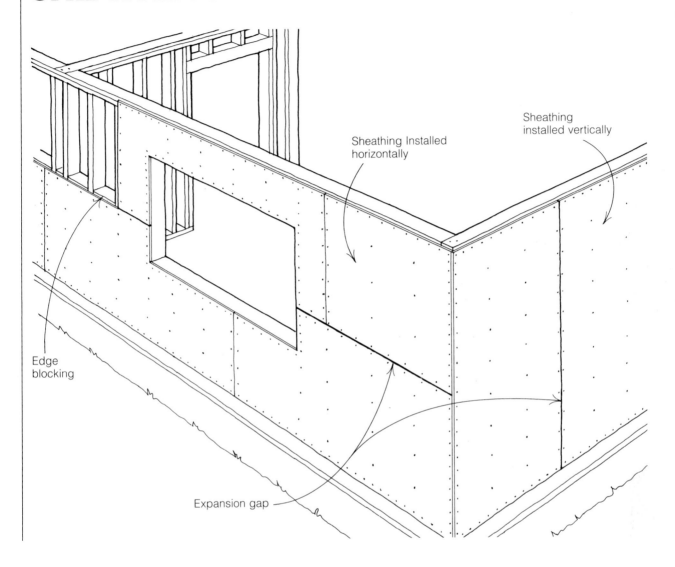

Sheathing installed vertically

Sheathing Installed horizontally

Edge blocking

Expansion gap

Sheathing can serve several different functions. It can provide both lateral strength, thus eliminating the need for wall braces, and vertical strength. It helps to keep cold out of the house and to strengthen walls facing high winds. On houses that are covered with vertical siding, wall sheathing can provide a nailing surface. It can also help keep batts of insulation in place. In some areas, sheathing is nailed only on the corners to hold the building plumb, and then the rest of the building is covered with insulation board.

The most common sheathing materials are square-edged plywood and oriented strand board (OSB). Walls covered with these materials are called shear walls; they may be required on both exterior and interior walls. In the West, where the ground gets a little nervous at times, the code often requires the exterior of the entire first floor to be sheathed on two-story wood-frame buildings, and second and third stories on taller ones. Some buildings, such as those built on hillsides, may need input from a structural engineer to determine which walls need to be sheathed.

In some areas of the country, walls are sheathed before they are raised. This is not done much in the West, where houses are frequently built on slabs, because the sheathing would interfere with the plumbing pipes that have already been installed in the floor. Also, walls should be perfectly square before they are sheathed, and squareness is easier to achieve after the walls are raised. Finally, sheathed walls are heavier to raise than unsheathed walls. All in all, it seems easier to sheathe walls after they've been raised, plumbed and lined.

If the walls were built with standard precut studs, a single sheet of 8-ft. sheathing will cover the framing from bottom to top plates. On taller walls, a row of blocks can be nailed into the frame at the 8-ft. height to provide a nailing surface for the end joints, or you can use 4x10 sheets of sheathing.

Check your plans and the code to be sure what grade of sheathing is called for. Shear walls often require a better grade than is used on floors. A commonly used material is CDX plywood, ⅜ in. thick on walls with studs at 16 in. on center and ½ in. thick on walls with studs at 24 in. on center. If sheathing panels are being used as backing for shingle siding,

they should be at least ⅝ in. thick so that the nails will hold better.

Wall sheathing can be installed horizontally or vertically. Horizontal sheets may not provide as much shear strength as vertical sheets unless they are blocked and nailed along their edges. Vertical sheets don't have to be staggered. Start at a corner and drive two 16d nails between the bottom plate and the floor sheathing or the concrete slab, and let them protrude an inch or so to set the panel on for support. Square the panel with the corner and tack it in place.

If a sheathing panel doesn't break on the center of a stud, it can be ripped to fit or an extra stud can be nailed in the wall to provide adequate backing. It is often more efficient to sheathe right over doors and windows. Then, when the whole wall has been sheathed, you can go back with a chainsaw or reciprocating saw and cut out the openings.

Nailing

Sheathing used to windproof a building requires fewer nails than sheathing used to stabilize a building in earthquake country. A typical code requirement is 8d nails at 4-6-12, just as with floor sheathing (see pp. 61-62). But check the nailing schedule on the plans to be sure.

Pneumatic nailers are great for nailing off shear walls. They are much faster than hammers. If you're using a pneumatic nailer, make sure that the nails are driven flush with the surface of the sheathing. If they're driven any deeper they lose a significant amount of their holding power. Also, don't hold the nailer directly in front of your face. If a nail should hit a metal strap, such as a metal brace beneath the sheathing, it could drive the nailer straight back into your front teeth.

Sheathing interior walls

At times, sheathing is required on interior walls to add to a building's structural stability. This is especially true in sections of a building where there is not enough wall space for let-in braces. Panels nailed on short walls, as on the back side of a closet or alongside a garage-door opening, can help stabilize an entire section of a house.

Nailing Schedule

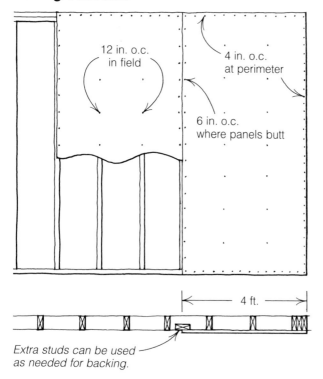

12 in. o.c. in field

4 in. o.c. at perimeter

6 in. o.c. where panels butt

4 ft.

Extra studs can be used as needed for backing.

FRAMING CEILINGS

Joisting for a
Gable Roof

Joisting for a
Hip Roof

4

JOISTING FOR A GABLE ROOF

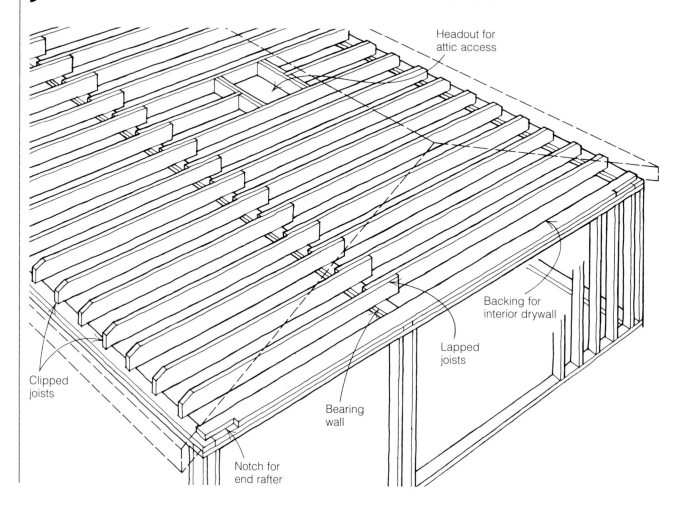

Headout for attic access

Backing for interior drywall

Lapped joists

Bearing wall

Clipped joists

Notch for end rafter

Ceiling joists can be nailed to the tops of the walls once the walls have been plumbed and lined. These joists are an important structural part of a building. Once nailed into the plates, they help tie the building together. On a two-story house, the ceiling joists over the first floor serve as the floor joists for the second floor (for more on floor joists, see pp. 41-54). On a one-story house, joists form the ceiling to which drywall can be attached. They tie into the roof rafters at the plate line, preventing the weight of the roof from pushing the exterior walls out, causing the ridge to sag. In the attic, spaces between the joists can be filled with insulation.

Joist size

The size, spacing and direction of ceiling joists are given on the floor plan. The framer's job is to take these instructions and locate the joists correctly on the walls. The size of lumber needed for ceiling joists is determined by the load they will support, their spacing and span and the type of wood used (see the table on the facing page). Ceiling joists that won't be carrying a floor load do not need to be as large or closely spaced as typical floor joists. If the attic is going to be used for living space or storage, the ceiling joists will have to be beefed up to carry the added weight. Many building departments have charts available showing what size joist is needed under different conditions.

TYPICAL SPANS FOR CEILING JOISTS

Size of joists	Spacing of joists	Species: Douglas Fir-Larch Grade: No. 2 or better		Species: Hem-Fir Grade: No. 2 or better	
		Drywall ceiling	Plaster ceiling	Drywall ceiling	Plaster ceiling
2x4	12 in.	12 ft. 8 in.	11 ft. 0 in.	11 ft. 10 in.	10 ft. 4 in.
	16 in.	11 ft. 6 in.	10 ft. 0 in.	10 ft. 9 in.	9 ft. 4 in.
	24 in.	9 ft. 11 in.	8 ft. 9 in.	8 ft. 10 in.	8 ft. 2 in.
2x6	12 in.	19 ft. 11 in.	17 ft. 4 in.	18 ft. 4 in.	16 ft. 3 in.
	16 in.	17 ft. 9 in.	15 ft. 9 in.	15 ft. 10 in.	14 ft. 9 in.
	24 in.	14 ft. 5 in.	13 ft. 9 in.	12 ft. 11 in.	12 ft. 11 in.
2x8	12 in.	26 ft. 2 in.	22 ft. 10 in.	24 ft. 2 in.	21 ft. 5 in.
	16 in.	23 ft. 5 in.	20 ft. 9 in.	20 ft. 11 in.	19 ft. 5 in.
	24 in.	19 ft. 0 in.	18 ft. 2 in.	17 ft. 1 in.	17 ft. 0 in.

Layout

The quickest way to run a layout for ceiling joists is to hop up onto the double top plate with a measuring tape and keel and start marking. With a bit of practice and a stout heart, you can learn to do this. It is easier to do when the walls are framed with 2x6s. If you're not ready to "walk the plates," the layout can be done from a ladder or sawhorse.

Starting with the exterior walls, hook a long tape on an outside corner and make a mark every 16 in. o.c. (or 24 in., as per plans), adding an X or dash to indicate which side of the layout line the joist will go. Ceiling joists are frequently laid out just like floor joists; they span from an exterior wall to an interior load-bearing wall, where they lap with another joist running to the opposite wall. Unless your code or the plans call for blocks where the joists lap, run a layout on the interior wall also. Any change of direction in joist layout will be noted on the plans. Once the layout is done, scatter the joists flat across the double top plates.

Clipping and cutting joists

The ceiling joists will tie into the roof rafters. Sometimes the ends of the joists will stick above the slope of the roof unless they are cut, or "clipped." This situation occurs most often when joists are made from stock that is larger than the rafters. There are sever-

Clipping Ceiling Joists

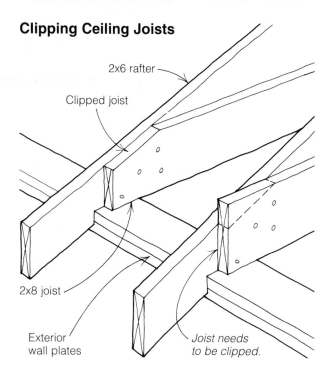

2x6 rafter

Clipped joist

2x8 joist

Exterior wall plates

Joist needs to be clipped.

al ways to clip the joists. When the clip is substantial, many framers prefer to make a template, as shown in the sidebar at right, so that joists can be marked and cut right at the lumber pile or up on the walls after being scattered. When the clip is minor, you can wait until the joists and rafters are nailed in place and then cut them with a saw or a rigging ax.

Check the plans for the rafter-blocking style (see p. 147). If frieze blocking is to be installed flush with the exterior wall, the ceiling joists should be held back 1½ in. from the outside edge of the wall, allowing 2 in. of bearing on the double top plate of a 2x4 wall. If the plans call for frieze blocks to be installed outside the exterior wall, the joists will have full bearing on the plate.

Once the joists have been scattered, you can use the building to do the rest of the measuring for you. Joists need to lap at least 4 in. over an interior wall. Any joist that extends more than 12 in. over an interior wall should be cut back. If joists need to be clipped, make sure the crown is up, then mark them with the template and cut. When joists span from one exterior wall to another, cut them exactly to length, flush with the outside of the building or set back to allow room for frieze blocks. These cuts can be made while walking on the plates and joists.

Making a template for clipping joists

Before you make a template for clipping joists, you should study pp. 138-144 on how to make a bird's-mouth cut in a rafter. Once you know something about cutting rafters and understand how they tie to joists at the plate line, the template is easy to make.

Take a piece of rafter stock, lay out the bird's mouth with an 8-in. or 10-in. rafter tail, make the saw cuts and set the rafter on an exterior wall. Next place a 12-in. to 16-in. piece of 1x8 the same width as the joist alongside the rafter. A mark made along the top of the rafter onto the 1x8 will indicate the excess material that needs to be clipped from the joist. Make the cut and then nail a short 6-in. piece of 1x2 to the edge so that the template can be used to mark all joists. To use the template, sight crown up on the joists, lay the template on the end, mark the clip and cut off the excess material.

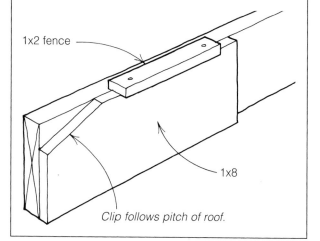

1x2 fence

1x8

Clip follows pitch of roof.

Gable-end joists

The first joist on a gable end can be set on edge just inside the wall (see drawing A on the facing page) to provide backing for drywall and leave room for nailing gable studs to the plate and the end rafter (for more on installing gable studs, see pp. 153-154). But many carpenters prefer a different method of joisting at the gable end. On a 2x4 wall, nail a 2x6 flat to the double plate, flush with the outside (draw-

Gable-End Joists

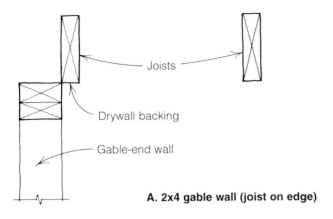

A. 2x4 gable wall (joist on edge)

Joists

Drywall backing

Gable-end wall

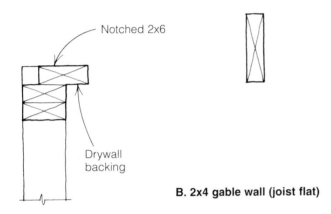

Notched 2x6

Drywall backing

B. 2x4 gable wall (joist flat)

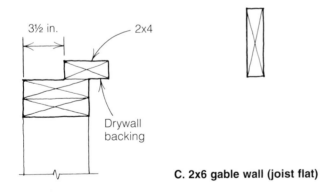

3½ in.

2x4

Drywall backing

C. 2x6 gable wall (joist flat)

ing B). This will give 2 in. of backing for drywall, and gable studs can be nailed on top of the 2x6. A notch will then have to be cut on the outside end of the 2x6 to allow room for the end rafter to rest on the plate (see the drawing on p. 128). The length of the notch will depend on the pitch of the roof. On a 2x6 wall, nail a 2x4 flat, 3½ in. from the out-

Ceiling joists should be toenailed to the plates with one 16d nail on each side.

side edge (drawing C). This placement will leave a 1½ in. overhang for drywall and allow gable studs to be nailed to the double top plate.

Rolling and nailing joists

With the crown side up, start nailing the joists on edge with a 16d nail in each side. Check the plans or code to see if blocking is required at the laps; often it is not required in an attic (for more on blocking between lapped joists, see pp. 46-48). Lapped joists do have to be nailed together with two 16d nails and then toenailed to the plates with one 16d on each side. Any joist that passes over an intermediate wall is secured to the plate with a 16d nail on each side. This helps to hold these walls plumb and straight and also strengthens the entire structure. Install backing for drywall on parallel walls (see p. 54).

Headouts in ceilings

At least one headout is usually necessary in the ceiling joists to allow for access to the attic. Most codes require this access to be at least 30 in. by 30 in. It is best to try to locate the headout out of sight, such as in a closet, rather than spoil the appearance of a larger room. The headout can be constructed like those that were framed into the floor joists (see pp. 49-50). If the ceiling won't be carrying extra weight, the cut-off joists can be nailed onto the header joists using blocks (sometimes called "pressure blocks") rather

Stub-Joist Framing

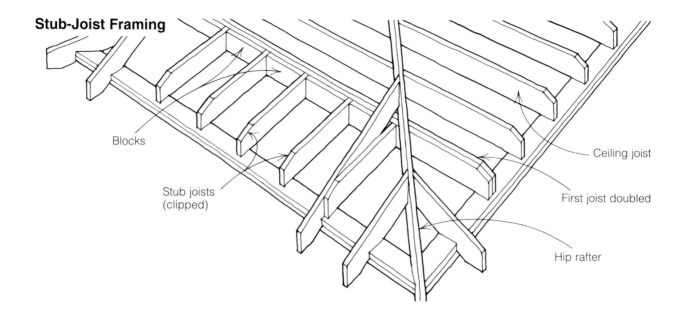

Blocks

Stub joists
(clipped)

Ceiling joist

First joist doubled

Hip rafter

pass over the top of it. The next step is to install the rafters and nail frieze blocks between them. Nail flat blocks for drywall backing to the plate behind the frieze blocks. Then cut strongbacks from scrap 2x stock and run them flatwise from the backing block at the plate line, across the flat joist, to the first on-edge joist. You won't need many strongbacks; spacing them about 4 ft. on center should do.

Secure the strongbacks to the backing blocks and to the upright joist with a couple of 16d nails at each end. At the joist end the strongbacks must be held up 1½ in. from the bottom edge to remain level. Finally, pull the flat joist up to the strongbacks and secure them with three or four more nails, angling the nails slightly for better holding power. The strongbacks will stiffen and support the flatwise joist. With this step complete, the ceiling is joisted and ready for drywall.

On steeper pitched roofs, the joist closest to the outside plates can often be installed the standard 16 in. out from the wall. Then, nail in the flat backing blocks behind the frieze blocks at the plate line. If this last joist can't pass under the hip or valley rafter, run it flat or head it off (as shown in the drawing at right). If you head it off, additional blocks can be nailed on top of the plates once the roof is stacked to create backing for drywall around the hip or valley.

Heading Out the Joists

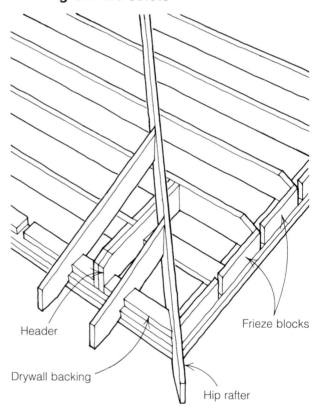

Header

Drywall backing

Frieze blocks

Hip rafter

Counting Common Rafters

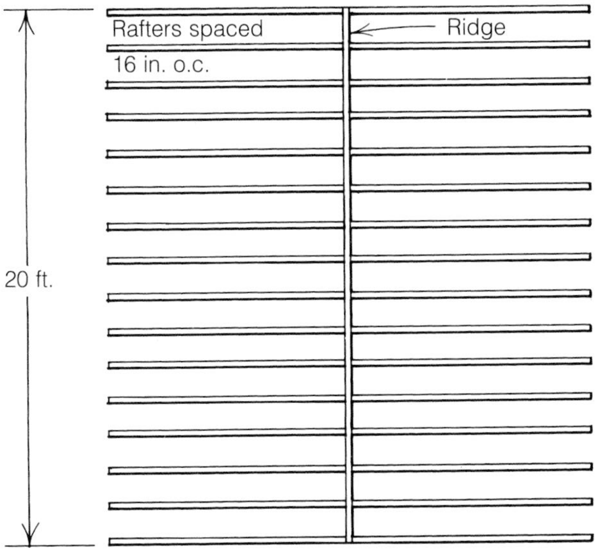

To calculate the number of common rafters needed for a 16-in. o.c. gable roof, divide the ridge length by 4, multiply the result by 3, then add 1. Multiply this result by 2 to cover both sides. (20 ÷ 4 x 3) + 1=16; 16 x 2 = 32

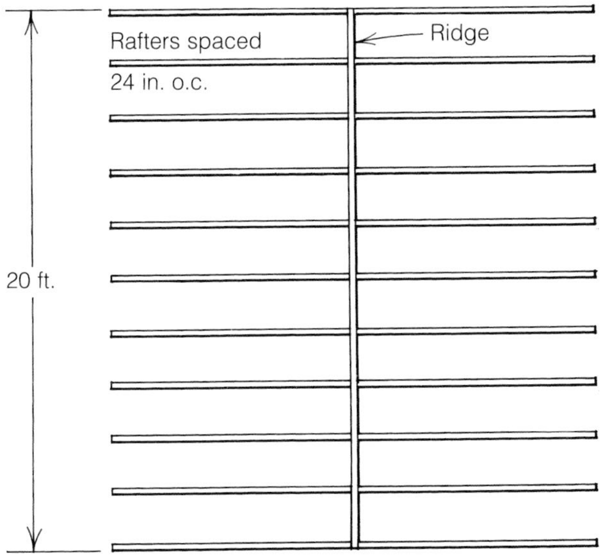

If the rafters are spaced 24 in. o.c., the number of common rafters needed is the length of the ridge in feet plus 2.

most efficient to mark and cut all the rafters at once. To do this, you need to build a couple of simple aids, as described in the sidebars on pp. 140-141. The first device is a pair of heavy-duty rafter horses, which need to be strong enough to hold a whole rack of rafters. The other tool is a rafter template (also called a "layout tee"), for scribing the ridge cut and bird's-

(continued on p. 142)

COMMON-RAFTER TABLE
4-in-12 pitch (18½° angle)

Span Ft.	Length Ft.	Length In.	Span In.	Length In.
1		6⅜	¼	⅛
2	1.	0⅝	½	¼
3	1.	7	¾	⅜
4	2.	1¼	1	½
5	2.	7⅝	1¼	⅝
6	3.	2	1½	¾
7	3.	8¼	1¾	⅞
8	4.	2⅝	2	1
9	4.	8⅞	2¼	1⅛
10	5.	3¼	2½	1⅜
11	5.	9⅝	2¾	1½
12	6.	3⅞	3	1⅝
13	6.	10¼	3¼	1¾
14	7.	4½	3½	1⅞
15	7.	10⅞	3¾	2
16	8.	5¼	4	2⅛
17	8.	11½	4¼	2¼
18	9.	5⅞	4½	2⅜
19	10.	0⅛	4¾	2½
20	10.	6½	5	2⅝
21	11.	0⅞	5¼	2¾
22	11.	7⅛	5½	2⅞
23	12.	1½	5¾	3
24	12.	7¾	6	3⅛
25	13.	2⅛	6¼	3¼
26	13.	8½	6½	3⅜
27	14.	2¾	6¾	3½
28	14.	9⅛	7	3¾
29	15.	3⅜	7¼	3⅞
30	15.	9¾	7½	4
31	16.	4	7¾	4⅛
32	16.	10⅜	8	4¼
33	17.	4¾	8¼	4⅜
34	17.	11	8½	4½
35	18.	5⅜	8¾	4⅝
36	18.	11⅝	9	4¾
37	19.	6	9¼	4⅞
38	20.	0⅜	9½	5
39	20.	6⅝	9¾	5⅛
40	21.	1	10	5¼
41	21.	7¼	10¼	5⅜
42	22.	1⅝	10½	5½
43	22.	8	10¾	5⅝
44	23.	2¼	11	5¾
45	23.	8⅝	11¼	5⅞
46	24.	3	11½	6
47	24.	9¼	11¾	6¼
48	25.	3⅝	12	6⅜
49	25.	9⅞		
50	26.	4¼		

Making a rafter template

Begin with a 2-ft. long 1x that is the same width as the rafters. Use a small rafter square or a framing square to scribe the ridge plumb cut at one end of the template. To scribe the ridge cut with a small rafter square, place the square on the edge of the 1x, rotate it at the pivot point to the correct roof-pitch number (4 for a 4-in-12 pitch) on the "common" line and mark along the pivot side for the cut. Move down the template about 1 ft. and scribe the heel cut (another plumb cut) of the bird's mouth, extending this line across the top edge of the template. This will serve as your registration or guide mark when laying out the bird's mouth on the rafter stock.

To scribe the ridge plumb cut with a framing square, hold the unit rise (4 in.) of the tongue on the edge of the 1x and the unit run (12 in.) of the blade on the same edge. Scribe along the tongue side for the plumb cut. Move the square down the template and scribe a second plumb mark for the heel plumb cut of the bird's mouth.

To mark the level seat cut of the bird's mouth it is important to understand the concept of height above plate (HAP). HAP indicates the amount of stock that is left between the seat cut of the bird's mouth and the top edge of the rafter. This amount varies depending on the width of the rafter stock, the steepness of the pitch and the length of the seat cut. The basic rule is to leave enough stock above the seat cut of the bird's mouth so as not to weaken the rafter tail.

The only time you usually have to worry about HAP is when you are using 2x4 rafter stock. In this case, measure 2½ in. down from the top edge of the template on the heel plumb-cut line. At this point, using the small square, scribe a line for the level part of the bird's mouth perpendicular to the plumb cut. Leaving at least 2½ in. HAP ensures a strong rafter tail on 2x4 stock, even if the seat cut does not completely cover the plate.

This normally presents no problems structurally as long as the rafters are stacked, nailed and blocked properly. (Many building codes only require a minimum bearing on the plate of 1½ in.) With 2x6 or larger rafters, the seat cuts can be 3½ in. long (longer for 2x6 walls), without weakening the tails.

Cut the template carefully so the rafters will be marked accurately, and then nail a 1x2 fence to the upper edge. The fence allows you to place the template on each rafter and transfer the marks rapidly and accurately. Make sure you cut the fence short enough so that it won't prevent you from seeing the top edge of the ridge end or the registration mark for the bird's mouth.

Rafter Template

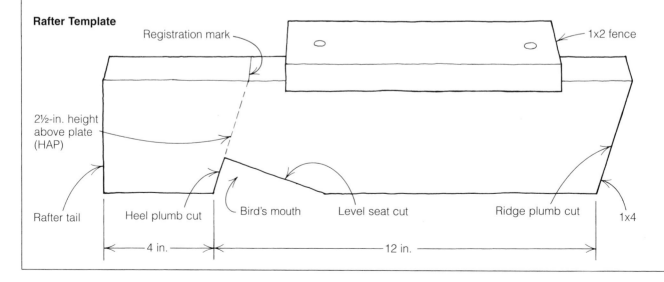

Registration mark

1x2 fence

2½-in. height above plate (HAP)

Rafter tail — Heel plumb cut — Bird's mouth — Level seat cut — Ridge plumb cut — 1x4

|← 4 in. →| |← 12 in. →|

Marking cut lines with a small rafter square

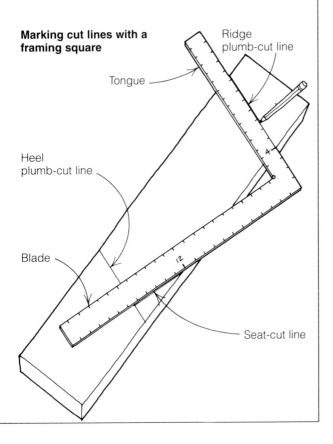

Pivot point

Ridge plumb-cut line

Heel plumb-cut line

Seat-cut line

90°

COMMON

Height above plate

Add for tail.

Marking cut lines with a framing square

Ridge plumb-cut line

Tongue

Heel plumb-cut line

Blade

4

12

Seat-cut line

Building rafter horses

There are a number of different ways to construct rafter horses. One method is to take four pieces of 2x6 about 3 ft. long, lay them flat and nail on a pair of blocks made from 2x stock. Leave a 1½-in. gap between the blocks so that a long 2x8 or 2x10 can be inserted on edge, on which the rafters will rest. A second method is to cut 1½-in. notches about 5 in. deep into short lengths of 4x12; then slip the long 2x on edge into these notches.

Either type of rafter horse can easily be carried from job to job, and will hold the rafters up high enough to provide plenty of clearance for cutting. Set the horses up on level ground near the lumber pile, so you can quickly load them with stock. If necessary, nail blocks on top of the horses at each end of the racked-up rafters to hold them stable.

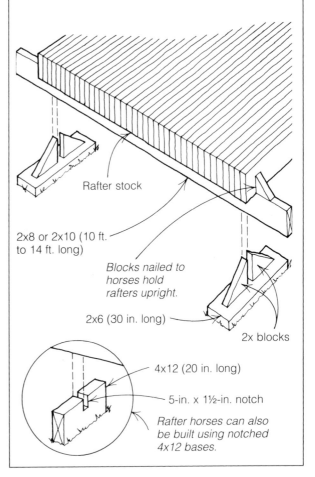

Rafter stock

2x8 or 2x10 (10 ft. to 14 ft. long)

Blocks nailed to horses hold rafters upright.

2x6 (30 in. long)

2x blocks

4x12 (20 in. long)

5-in. x 1½-in. notch

Rafter horses can also be built using notched 4x12 bases.

Cuts on a Common Rafter

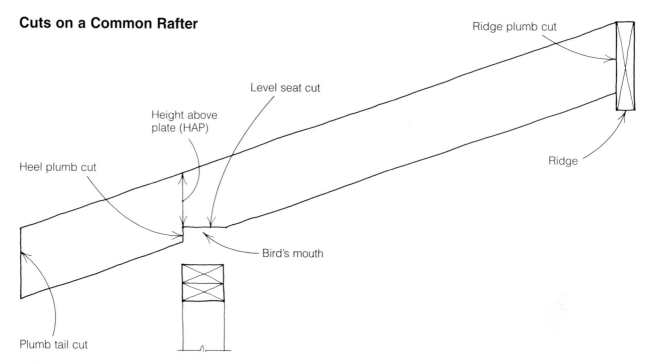

Ridge plumb cut

Level seat cut

Height above
plate (HAP)

Heel plumb cut

Ridge

Bird's mouth

Plumb tail cut

Simple site-built rafter horses allow you to mark and cut all of the common rafters at once.

mouth cuts. The ridge cut is a plumb cut on the rafter stock that will fit against the ridge when the roof is built. The bird's mouth is the notch in the rafter that rests on the double top plate. It consists of a plumb heel cut and a level seat cut.

Rafters are generally cut using a standard 7¼-in. circular saw. This saw isn't the first choice for production roof cutters, who prefer to use more spe-

cialized tools (especially when cutting simple gable roofs), but it is the more affordable choice for most all-purpose carpenters. If you are using a standard saw, begin by loading the rafter stock on edge on the horses with their crowns up, just as they will be when installed. Leave both the top and bottom ends of the stock hanging over the horses about 1 ft. Make sure that the rafter stock is long enough to include the tail or eave section of the rafter.

The rafters should be flush on the ridge end. An easy way to flush up the rafters is to hold the face of a stud against the end of the rafters and pull them against it one at a time using your hammer claws. From the flush end measure down on the two outside rafters and mark the theoretical length of the rafters, subtracting half the ridge thickness if this has not been done. Snap a chalkline across the tops of the rafters to connect the marks. This marks the location of the plumb heel cut on the bird's mouth.

Next, place the rafter template on the first rafter, flush with the ridge end, and scribe the ridge-cut line. Slide this rafter over and mark the second one, and so on down the line, leaving all the rafters on edge. When all the ridge-cut lines have been marked, align the registration mark on the template with the chalkline on the rafters and begin marking the bird's-mouth cut lines.

Holding an aligning board in place with your knees, pull the rafters flush using the hammer claws.

With the ends flush and the rafter lengths marked, scribe the ridge-cut line using the template.

Use the same template to mark the bird's mouth. Match the registration mark on the template with the chalkline on the top edge of the rafter.

Laying Out Rafters

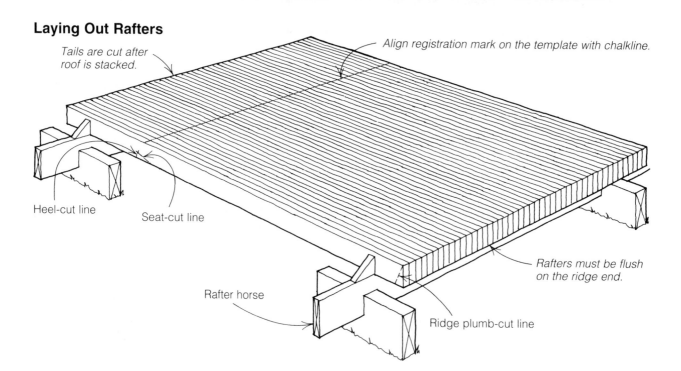

Tails are cut after roof is stacked.

Align registration mark on the template with chalkline.

Heel-cut line

Seat-cut line

Rafters must be flush on the ridge end.

Rafter horse

Ridge plumb-cut line

Once all the rafters have been marked, make all the ridge cuts.

Some buildings have barge rafters that form an overhang at the gable ends supported by lookouts (see pp. 155-156). A lookout is a 2x4 laid flat that butts against the first inboard rafter, passing through a notch cut in the end rafter and cantilevering out to support the barge. Lookouts are usually installed near the ridge, just above the plate line and 32 in. on center in between (closer for wide overhangs or heavy barge rafters). If the overhang is to be sheathed with 4x8 panels and left exposed, space the lookouts every 48 in. on center from the bottom. This way they will hide the joints of the sheathing. Pick out four straight rafters, stack them together on edge and lay out notches 3½ in. wide.

When all the rafters have been marked, make all the ridge cuts with your circular saw, moving the rafters over one at a time. Then flip the rafters onto their sides and make the cuts for the bird's mouth, overcutting just enough to remove the wedge but not so much that you weaken the tail section. Cut the notches for the lookouts by first making two square crosscuts 1½ in. deep, 3½ in. apart, across the top edges of the four end rafters. Then turn the rafters on their sides and plunge-cut the bottom of the notch, removing just enough wood for a 2x4 to fit snug.

If lookouts are needed to help carry the barge rafter, now is the time to mark and notch the end rafters.

Production layout and cutting

People who cut a lot of rafters, especially professional roof cutters, use a variety of specialized tools that allow them to gang-cut common rafters, saving time in the process. These production methods require that the rafters be stacked on edge on the horses with the crowns facing *down*. Flush up one end and snap a chalkline about 3 in. down from the flush end (the greater the roof pitch and rafter width, the greater this distance needs to be). The chalkline marks the short point of the ridge plumb cut. Measure down from this line the theoretical rafter length and snap another chalkline, which marks the heel cut of the bird's mouth. Then, measure back up from this mark about 2½ in. and snap a third line to mark the seat cut of the bird's mouth. This measurement will vary depending on the size of the rafters, the pitch of the roof and the cutting capacity of your saw (more on this later).

The rafters are now ready to be gang-cut. A 16-in. Makita beamsaw will cut through a 2x4 on edge at more than an 8-in-12 pitch (33¾°) and will saw

A 16-in. circular saw (top) or a worm-drive saw equipped with a beam-cutting attachment (above) can make the ridge plumb cut on a rack of rafters in a single pass.

A circular saw fitted with a swing-table accessory (left) can make the seat cut on a rack of rafters in a single pass. The more expensive dado rig (right) can gang-cut the entire bird's mouth in one quick pass.

most of the way through a 2x6 at a 4-in-12 pitch (18½°). To determine the angle at which to set your saw, check a rafter-table book.

For steeper pitches or wider stock, make a single pass down the chalkline with the beamsaw or standard circular saw, and then finish each cut with the circular saw, moving the rafters over one at a time. This way, the only mark needed is the chalkline. The kerf from the first cut acts as a guide for the second cut. To make this process go even faster, apply paraffin to the sawblade and table. Also, try to stay close to your power source. If you have to roll out 100 ft. of cord or more, the saw will lose some power and won't operate at its maximum efficiency.

Another method for gang-cutting ridges is to use a chainsaw-type beam cutter such as the Linear Link model VCS-12 saw (Muskegon Power Tools, 2357 Whitehall Rd., Muskegon, Mich. 49445). The VCS-12 is a worm-drive saw fitted with a bar and cutting chain that can cut to a depth of 12 in. at 90° and is adjustable to cut angles up to 45°. The saw leaves a good clean ridge cut and is easy to use, although it does require special care to use it safely. Read the in-

struction manual carefully. Also available is a beam-cutter conversion kit for worm-drive saws (Prazi USA, 118 Long Pond Rd., #G, Plymouth, MA 02360).

With the right tools, the bird's mouths can also be gang-cut with the rafters on edge. For the heel cuts, set your circular saw to the same angle as the ridge cut and to the proper depth, and then make a single cut across all the rafters. Seat cuts are made using a circular saw fitted with a swing table (Big Foot Saws, Box 92244, Henderson, NV 89015). A swing table replaces the saw's standard table and allows the saw to be tilted to angles up to 75°. Just set the swing table to 90° minus the angle of the plumb cut and make the cut in one pass.

One problem with using a swing table is that it won't give you a deep cut at this sharp angle, so it limits the amount of bearing that the rafters will have on the top plates. This shallow cut is of little concern if the roof is framed properly. But for jobs requiring a deeper cut, Pairis Enterprises also makes a swing table to fit the 16-in. Makita beamsaw.

Gang-cutting bird's mouths works especially well because you can avoid overcutting on the heel and seat cuts. Once you get used to working with these specialized tools, you'll find that it takes longer to stack the rafters on the horses than it does to cut them.

An even faster way to make seat cuts is to use an 8¼-in. worm-drive saw equipped with a universal dado kit, a rig that has been around for over 15 years. This kit (also manufactured by Pairis Enterprises) consists of an accessory arbor extension that allows the saw to accept a stack of carbide blades up to 3¼ in. wide. With this setup, the bird's mouths can be gang-cut in a single pass and require just one chalkline for the heel cut. The dado setup is easy to control as long as it's used for its intended purpose, which is to plow out stock on a horizontal surface. It will make a noise somewhat like a router and kick pieces of stock out the front end. Though the guard should prevent wood chips from hitting you in the face, you should wear safety glasses and ear protection when running a saw with a dado rig. The one drawback to the rig is its cost—about $750 including the saw—but if you cut a lot of roofs, it will quickly pay for itself.

When all the rafters have been cut, carry them to the house and lean them against the walls, ridge end up. Rafter tails will be cut to length once the roof is stacked.

Frieze blocking

Frieze blocks are cut from 2x scrap lumber and nailed between the rafters at the wall plates. They are not always required by code, but they stabilize the rafters and provide perimeter nailing for roof sheathing. Frieze blocks can easily be drilled and screened for attic vents. The blocks are cut 14½ in. long for rafters set at 16-in. on center and 22½ in. long for rafters at 24-in. on center. Once cut, the blocks can be stacked on a small piece of plywood placed on the joists, or hung from the top plate by a 16d nail driven in each block.

There are two common ways to block a gable roof. The first is to install the blocking flush with the wall, allowing the blocks to serve as backing for the exterior siding. This method requires the blocks to be ripped so they won't stick up above the roof

Two Ways to Install Frieze Blocks

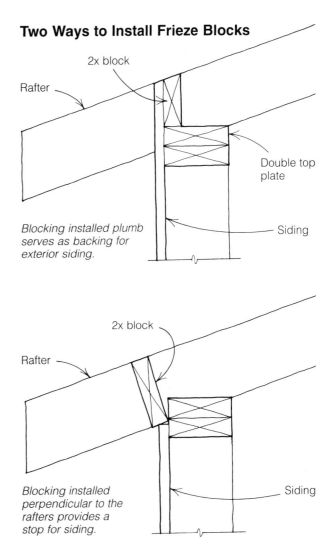

Blocking installed plumb serves as backing for exterior siding.

Blocking installed perpendicular to the rafters provides a stop for siding.

line. The other method is to install the blocking perpendicular to the rafters just outside the plate line. This method of blocking provides a stop for the siding, eliminating the need for it to extend between the rafters (see the drawing above). It also eliminates the need to rip the blocking. With either method, the blocks should be installed when you nail on the rafters.

Staging and ridge layout

Now it's time to prepare a sturdy platform from which to nail the rafters to the ridge. The easiest way to do this is to tack 1x6s or plywood strips across the joists under the ridge line. Make this catwalk about 2 ft. wide running the full length of the building. If the ridge is too high to reach comfort-

The few minutes it takes to nail down a catwalk under the ridge makes installing rafters much easier and safer.

ably from the catwalk, you need to set up staging on the ceiling joists. Lay down one or two rows of plywood directly under the ridge. Then set two strong sawhorses on the plywood and run 2x10 or 2x12 planks between them. Experienced carpenters facing a high ridge often frame and brace the bare bones of the roof from the catwalk and then install the rest of the rafters while walking on the ridge. In a room with a cathedral ceiling, you will have to set up staging on the floor. Common sense dictates that any scaffold you use be safe (see the sidebar on the facing page).

Next, lay out the ridge board. Most codes require that the ridge be one size larger than the rafters to ensure proper bearing (for example, 2x4 rafters need a 2x6 ridge). Most roofs will require more than one piece of ridge stock to span the length of the building. As you lay the ridge boards out, cut them at a layout mark so that each joint falls in the center of a rafter pair. The rafters will then help to hold the ridge together. Let the last ridge board run long—it will be cut to length after the roof is assembled. If the plans indicate that the ridge should run beyond the building line to help carry barge rafters, rip the overhanging piece to match the size of the rafters for a better appearance.

Be sure to align the layout of the ridge to that of the joists so that the rafters and joists will tie together at the plate line. If both are spaced 16 in. on center, every rafter will tie into a joist (see the drawing on the facing page); if the joists are 16 in. on center and the rafters 24 in. on center, then a rafter

Laying Out the Ridge

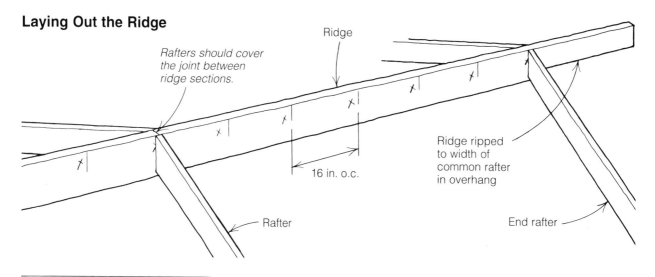

Rafters should cover the joint between ridge sections.

Ridge

16 in. o.c.

Ridge ripped to width of common rafter in overhang

Rafter

End rafter

will tie into every fourth joist. In either case, no layout is necessary on the top plates. Rafters will either fall next to a joist or be spaced the proper distance apart by frieze blocks nailed between them.

Nailing common rafters

Roof rafters can easily be installed (or "stacked") by two carpenters. First, pull up a straight rafter at the gable end. While one person holds the rafter at the ridge, the other toenails the bottom end to the double top plate with two 16d nails on one side and one on the other. The process is repeated with the opposing rafter. The two rafters will rest against each other temporarily, unless you're framing in a Wyoming wind. If that's the case, nail a temporary 1x brace diagonally from the rafters to a joist or plate.

Next, move to the opposite end of the first ridge section and toenail another rafter pair in the same way. Now pull the ridge up between the two pairs of rafters. There is no need to calculate the ridge

Tying Rafters to Ceiling Joists

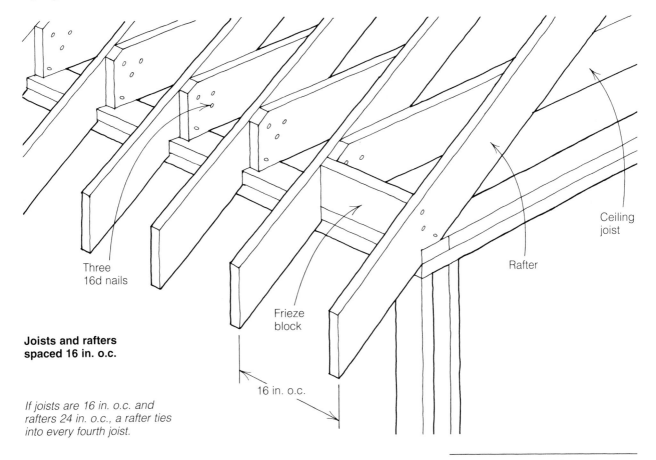

Joists and rafters spaced 16 in. o.c.

If joists are 16 in. o.c. and rafters 24 in. o.c., a rafter ties into every fourth joist.

Three 16d nails

Frieze block

16 in. o.c.

Ceiling joist

Rafter

The first rafters are installed at the ends of each ridge section. They are toenailed at the bottom to the double top plate. The ridge is then pulled up between the rafters and nailed on.

height—it is determined by the rafters without any measuring. Drive two 16d nails straight through the ridge into the end of the first rafter, then angle two more through the ridge into the opposing rafter. To keep from dulling a sawblade later when you sheathe the roof, avoid nailing into the top edge of the rafters.

Next, nail 2x4 legs from the top plate or joists to the ridge at both ends and one in the middle to give it extra support. If these legs need to be cut to different lengths to fit beneath the ridge, it means that the walls probably aren't parallel and, consequently, that the ridge board isn't level. In this case, pull the nails out of the rafter pair at the top plate on the high end of the ridge and slide the rafters out until the ridge rests on a leg. The small gap that results at the bird's mouth will be covered by siding or frieze blocks. The best way to avoid having to adjust the ridge board, of course, is to make sure that the walls are framed accurately in the first place.

Plumbing the ridge

The ridge can be plumbed in a couple of different ways. One way is to nail a straight 2x4 against the end wall, extending up to ridge height. Push the end rafters against the upright and install a 2x4 sway brace extending from the top plate to the ridge at a 45° angle, as shown in the top photo on the facing page. This is a permanent brace. Nail it in between the layout lines at the ridge so that it won't be in the way of a rafter. A second method to plumb the ridge is to use your eye as a guide. Sighting down from the end of the ridge, align the outboard face of the end rafters with the outside edges of the top and bottom plates, and then nail in a sway brace. Either way, the ridge can be plumbed without using a level, which means carrying one less tool up with you when you stack the roof.

With the rafters nailed on each end of the ridge, the ridge is supported by 2x4 legs. The vertical 2x4 at the far right is flush with the end of the wall. Plumb the ridge by pushing it against this 2x4 and then nailing on a permanent sway brace at 45°.

Raise the remaining ridge sections in the same way, installing the minimum number of rafter pairs and support legs to hold them in place. When you reach the far end of the building, eyeball the last rafter pair plumb with the wall, scribe the end cut on the ridge, slide the rafters over a bit and cut the ridge to length. Then reposition the rafters and nail them to the ridge. Install another sway brace at this end. Codes generally call for a sway brace on each end and one every 25 ft. On occasion, the ridge will extend into the overhang and be ripped to the size of the common rafters (see p. 148).

The rest of the rafters will now stack easily and make you feel that you're cruising right along. Nail through the sides of the rafters into the ends of the frieze blocks, using two 16d nails for up to 2x12 stock. Where a rafter falls next to a ceiling joist, drive three 16d nails through the rafter into the joist and a toenail down into the plate (see the drawing on p. 149). This forms a rigid triangle that helps tie the roof system together.

With the ridge plumb and braced, nail on the rest of the rafters at the ridge. Do one side, then the other. The bottoms are tied into ceiling joists and wall plates.

These rafters are resting on a 6x16 beam, which will be exposed in the cathedral ceiling.

Tying a High Rake Wall to End Rafters

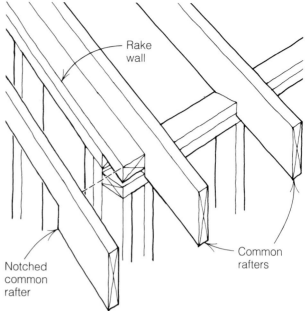

A rake wall that is high can be tied to end rafters by toenailing a notched common rafter flush with the outside of the wall, then nailing another rafter on the inside of the wall.

Commons on a cathedral ceiling

Rake walls are framed to allow for an open or cathedral ceiling (see pp. 103-104). The ridge is often a beam that spans the room with the rafters resting on top of it. In this case, the common rafters don't need a ridge plumb cut, but they do need a bird's-mouth cut at the plate line and may need a bird's mouth at the beam, depending on the code. Each rafter laps and ties into the opposing rafter with two 16d nails. Excess material at the top of the rafters can be trimmed off either before or after they have been nailed in place. The rafters need to be nailed and blocked at both the plate and the ridge.

If the rake wall wasn't framed perfectly, the rafter that sits on top of it can be made to fit a high or low wall easily enough. Many builders like to run the rake wall about 1 in. high to ensure a good tie between the wall and the roof. If the rake wall is high, set a common rafter alongside it just on the inside running from the ridge to the plate. Holding a pencil on top of the plate, scribe a mark the length of the rafter and rip off this excess material. This rafter is toenailed to the top of the rake wall flush with the outside of the building. Then set another rafter against the rake wall on the inside and nail it to the wall with 16d nails every 16 in. on center. This creates a good tie between the wall and roof and gives backing for ceiling drywall.

If the rake wall is a little low, set a common rafter flush with the outside and toenail it to the wall. On the inside, set a 2x on edge on top of the rake wall and nail it to the rafter and to the wall. Then set another common on the inside and nail it to the 2x. This will make for a strong tie between the wall and the roof. The rest of the commons can now be stacked much like regular rafters.

Collar ties, rafter ties and purlins

Building codes sometimes require the use of collar ties and rafter ties to help tie the roof structure together and purlins to reduce the rafter span. Collar ties are installed horizontally on the upper third of opposing rafters. They are usually made of 1x4 or 1x6 stock, and are placed every 4 ft. and secured with five 8d nails on each end so that they help tie the opposing rafters together.

When the joists run at a right angle to the rafters, rafter ties are nailed to the rafters near the plate line

Collar Ties, Purlins and Rafter Ties

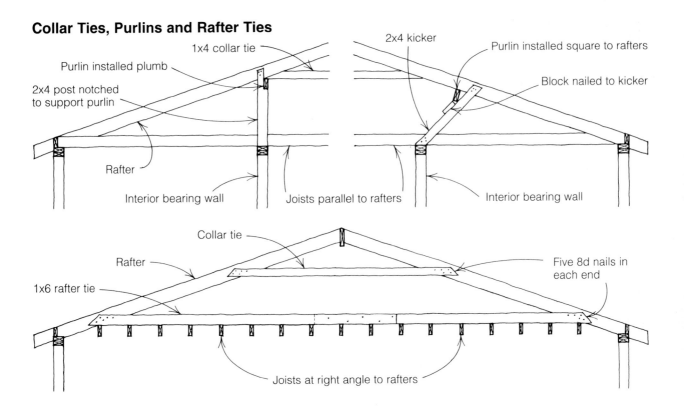

Purlin installed plumb

1x4 collar tie

2x4 post notched to support purlin

Rafter

Interior bearing wall

Joists parallel to rafters

2x4 kicker

Purlin installed square to rafters

Block nailed to kicker

Interior bearing wall

Collar tie

Rafter

1x6 rafter tie

Five 8d nails in each end

Joists at right angle to rafters

with five 8d nails at each end. Rafter ties form the bottom chord of the truss triangle, preventing the load on the roof from pushing down, bowing walls outward and creating a sag in the ridge. Rafter ties are usually made from 1x4 or 1x6 stock and are installed every 4 ft. If the roof is going to be covered with heavy tiles, 2x ties may be called for.

Purlins are required when rafter spans are long. They should be placed as close as possible to the middle of the rafter span and run the length of the building. Purlins can be toenailed to the rafters either plumb or square. If there's a load-bearing interior wall beneath the center of the rafter span, install the purlin plumb, directly over the wall, and support it with several 2x4 posts that bear on the double top plate of the wall. The 2x4s are notched so that they support the purlin and can be nailed to the sides of the rafters.

If there isn't a wall beneath the center of the span, toenail the purlin square to the rafters and install 2x4 kickers up from the nearest load-bearing parallel wall at an angle not exceeding 45°. A block nailed to each kicker below the purlin will help hold it in place. Kickers and posts are typically installed

every 4 ft. Larger purlins, such as 2x12s, require fewer supports. When several lengths are needed to form the purlin for a long roof, lap them at least 4 ft. and nail them together with five 16d nails.

In some parts of the country, rafters must be tied to the top plates or blocking, or even down to the studs, with framing anchors or hurricane ties for added protection against earthquakes or high winds. Check your local code.

Gable-end studs

Gable ends are filled in with gable studs spaced at 16 in. or 24 in. on center. If a gable vent is called for on the plans, place the two center studs 14 in. apart to allow room for the vent directly under the ridge. Measure the length of these studs to the long point, then calculate the common difference of the gable studs, that is, the difference in length between successive studs. Once you know the common difference, you can quickly determine the lengths of the remaining studs without measuring each one separately. A pocket calculator makes this task easy.

To calculate the common difference, divide the rise by the run and then multiply the result by the

To lay out gable studs, lay the stock flush against a bottom plate. Measure and mark the longest studs (one on each side of a gable end), then subtract the common difference and mark the next set, and so on. The studs then need to be cut at the angle of the roof pitch.

on-center spacing. For a 4-in-12 roof with gable studs spaced 16 in. o.c., the equation goes like this: $4 \div 12 \times 16 = 5.33$. In other words, the difference in length between successive studs is 5.33 in., or about $5\frac{3}{8}$ in. Another way to calculate the common difference is to divide the rise by three and add the result and the rise together ($4 \div 3 = 1.33 + 4 = 5.33$). When gable studs are 24 in. o.c., double the rise to find the common difference.

Cut all the gable studs at the same time. If there are two gable ends, two sets of gable studs will be needed to fill them in, so cut four of each length. This is a good time to use up some 2x4 scrap. Lay the pieces out on edge, flush on one end. Mark the length of the first set, then mark each successive set by subtracting the common difference. Don't bother cutting any studs shorter than 16 in.

For the angle cuts, set your saw at the angle of the roof pitch (18½° for a 4-in-12 roof). Nail the gable studs plumb using your eye as a gauge. There is no need to lay out the double top plate or align the gable studs with the wall studs below. Be careful not to force a crown into the end rafters when nailing in the gable studs.

When installing gable-end studs, nail through the rafter into the stud, rather than vice versa, to keep from putting a crown in the rafter. When you're working along the gable end, be sure to keep your body well balanced at all times.

Barge rafters

The next step is to install the barge rafters, if the plans call for them. These are rafters that hang outside the building line, creating an overhang on the gable end. Barge rafters are usually cut from the same size material as the fascia (see pp. 156-158), which they tie into at the corners. A barge rafter doesn't require a bird's-mouth cut because it doesn't land on the double top plate. For this reason it has to be supported by other means, usually by the ridge, fascia and roof sheathing. The ridge board should extend beyond the building line to the point where the two opposing barge rafters meet. The rafters butt together over the end of the ridge board and are face-nailed to it (as shown in the drawing at left). At the bottoms the barge rafters are mitered to fit the fascia boards, which also extend beyond the building line. The roof sheathing cantilevers out the same distance and is nailed to the tops of the barge rafters.

Sometimes barge rafters don't overhang the gable, but are held out only enough to cover the edge of the siding. In this case the barge rafters should be furred out from the building by a 1x. For example, if the barge rafter is a 2x8, nail a 1x6 flush with the top of the end common rafter, then face-nail the barge to it. Adding the furring strip leaves room for siding to slip up under the barge and leave a clean, trimmed-out finish.

Two Ways to Install Barge Rafters

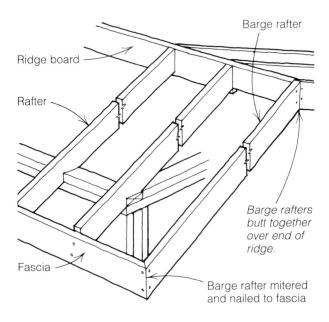

Ridge board

Rafter

Barge rafter

Barge rafters butt together over end of ridge.

Fascia

Barge rafter mitered and nailed to fascia

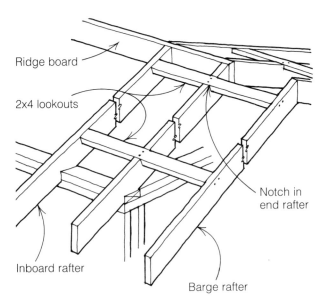

Ridge board

2x4 lookouts

Notch in end rafter

Inboard rafter

Barge rafter

The lookouts and ridge are cut to length after they've been nailed on. Snap a chalkline at the proper length and trim them off. Watch your step.

When barge rafters are supported by lookouts, nail the 2x4s flush with the top of the inboard rafter, then down into the notched end rafter, checking to make sure that this outside rafter sights straight. Next determine the width of the overhang from the plans, then snap a chalkline across the lookouts and cut them off with a circular saw. Face-nail the barge rafters to the ends of the lookouts with galvanized 16d nails. Since this framing will be exposed, make sure that the plumb cut at the ridge fits tightly and looks good.

Trimming rafter tails

Before the fascia can be installed, the rafter tails have to be marked and cut to length. Check the plans to determine the length of the overhang. Overhang is measured out from the wall, not down along the rafter. If, for example, the overhang is 20 in. and the fascia stock is a 2x (1½ in. thick), measure straight out from the building line 18½ in. and mark this point on the top edge of the rafters at both ends of the building. Snap a chalkline across the rafters, including the barges, connecting the marks.

If the tails are to be cut square, use a small rafter square to mark them. Most often they need a plumb cut, which can be marked using the template you made when cutting rafters (see the sidebar on pp. 140-141). You can also use a bevel square. If the tails are long and fragile, tack a 1x6 catwalk to them above the chalkline. Otherwise you can walk along the building line cutting off the tails with a circular saw. The barge rafters may need to be cut at a 45° angle to fit the fascia (see below).

Fascia and soffit

The fascia is a horizontal board that covers the ends of the rafters. If the plans call for a fascia, it is installed after the barge rafters have been nailed on. If the eaves will be closed in with a soffit, you may want to attach a subfascia, the same size as the rafters, to the rafter ends. Soffit sheathing can be nailed into this subfascia and to a 2x attached to the wall. The finish fascia will cover the subfascia and hide all the nails.

Another simple way to build a soffit is to cut a ¾-in. wide groove into the back side of the fascia just below the rafter tails. In 2x fascia, the groove

To ensure a straight fascia, mark and cut the rafter tails to length after the roof is stacked.

can be cut ½ in. deep with a router by nailing a 1x4 the length of the fascia to act as a guide. To close in the soffit, use ⅝-in. plywood that will slide into the groove in the fascia and nail into a 2x nailer attached to the wall.

The fascia does not have to be as strong structurally as framing lumber, so redwood, spruce, cedar and other species are often used. It can be roughsawn or smooth, as the design of the house requires, but it should be kiln dried and fairly free of knots. The pieces should be as long and straight as possible.

Stand the fascia boards around the building. If the roof pitch is not too steep, they can be pulled up and laid across the rafters. On steep pitches, drive some nails into the common rafters temporarily to hold the fascia stock in place. (For safety's sake, remember to take the nails out when they're no longer

Two Ways to Frame a Soffit

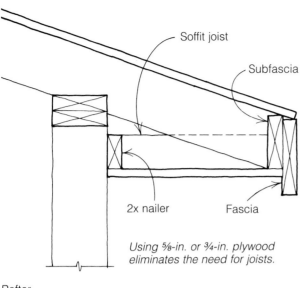

- Soffit joist
- Subfascia
- 2x nailer
- Fascia

Using ⅝-in. or ¾-in. plywood eliminates the need for joists.

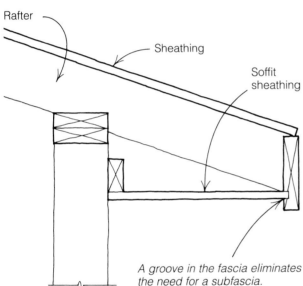

- Rafter
- Sheathing
- Soffit sheathing

A groove in the fascia eliminates the need for a subfascia.

Joining Fascia Board

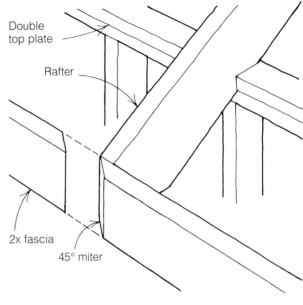

- Double top plate
- Rafter
- 2x fascia
- 45° miter

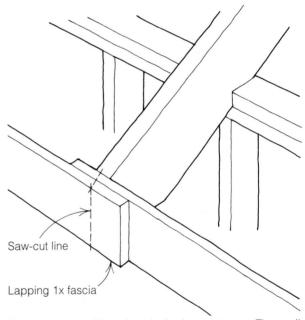

- Saw-cut line
- Lapping 1x fascia

Set your saw at 15° and cut both pieces at once. Then nail the mitered ends into the rafter.

needed.) Start at the barge rafter, which is generally of the same size and material as the fascia and ties into it with a 45° miter. The length of the barge was previously marked with a chalkline that also marked the length of all the commons. This mark is the short point of the 45° cut. Assuming the rafter is plumb, make the mark with the rafter template, set the saw table to 45° and make the cut.

Next make a miter cut on a fascia board, flushing this end with the barge-rafter cut. Cut a second miter on the other end of the fascia so that it will

break over a rafter end. Make the cut so that the miter face is exposed when the fascia is nailed on (as shown in the top drawing above). Cutting the miter this way allows the next piece of fascia to nail on top of the first piece rather than having to be shoved underneath.

Nail the fascia to the barge rafter before nailing it to the common rafters. Drive nails through both boards to create a tight joint.

Check how the miter cuts fit. Unless the building is absolutely square and the framing has been done perfectly, the joint in the fascia may be open slightly. A poorly fitting joint may also be caused by using a saw that isn't totally accurate. In either case, it may help to modify your saw a little. When you tip a sawblade over to 45° it is stopped in this position by a bolt hitting metal. Take a small round file and file out a bit of this metal so that you can reach 45½° or even 46°. Making this adjustment will ensure that the outside edge of corner miter joints will fit tightly.

Intermediate joints of 1x fascia can be cut differently, and more efficiently, guaranteeing a perfect joint every time. Nail the first fascia board to the rafter ends, letting it run wild. Then nail the second fascia board on so that it laps the first (see the bottom drawing at right on p. 157). Working from the roof, make sure that the piece of fascia to your right laps on top of the piece to your left. Flush the two boards together, mark square directly over a rafter end, and set your saw at 15° and deep enough to cut through both pieces at once. Nail the joint together for a perfect fit.

Most builders use regular hot-dipped galvanized nails to secure fascia. Stainless-steel nails are recommended near the coast to resist corrosion from salt air. Hold the fascia down on the rafters so that it won't interfere with the roof sheathing. The amount you drop the fascia varies with the pitch of the roof and the thickness of the fascia. Put a scrap of wood on the top edge of a rafter and let it extend beyond the plumb tail cut. Pull the fascia up to the piece of wood and nail it. Use your eye or a measuring tape to drop the fascia the same amount on each rafter. This way the sheathing can slide over the top of the fascia and be nailed into it.

Working with a partner holding up the other end, first nail the fascia to the barge rafter. Then drive two nails through the fascia into each rafter end, keeping the top nail high and the bottom nail low to ensure maximum holding power. If the fascia is roughsawn, an experienced carpenter can drive nails flush with the surface without leaving hammer marks, but if it is smooth, the nail can be driven close to the surface and then finished with a nail set.

Continue on around the building until the fascia is done. You've saved a lot of time by working from the roof rather than setting up scaffolding, which is especially time-consuming when the building has high gable ends or two stories. Take time to admire your work before you start sheathing. There is something about looking at a stacked roof that makes carpenters feel good about their work.

HIP ROOFS

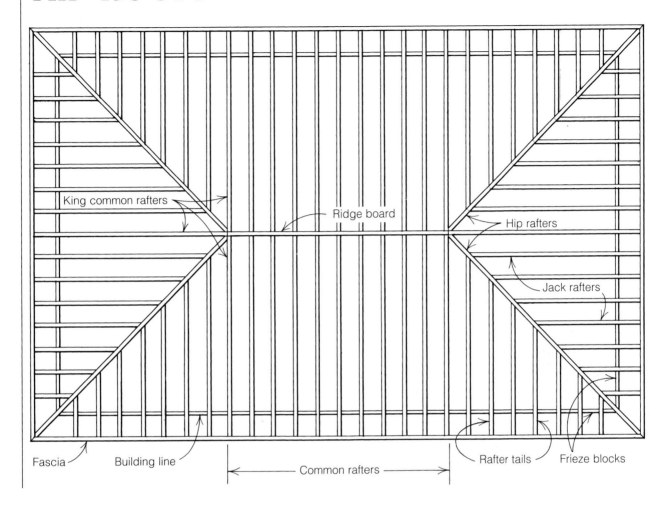

King common rafters

Ridge board

Hip rafters

Jack rafters

Fascia   Building line   Common rafters   Rafter tails   Frieze blocks

Once you have built a gable roof, the next step is to tackle a roof with hips. Whereas a gable roof has two sloping sides, a hip roof has four. It may look difficult, but with the roof-framing basics you learned in the previous section and the information presented here you should be able to lay out, cut and raise a simple hip roof.

A hip roof on a square building has four hip rafters extending at 45° angles from the outside corners of the building, meeting in the center. On a rectangular building, the hips come off the corners and tie into a level ridge. Like a gable roof, a hip roof has common rafters, but unlike a gable, it also has jack rafters, which run off the hip rafter. Jack rafters run parallel to, but are shorter than, the commons. Take a few moments to study the drawing above and familiarize yourself with the terms.

The lumber used for hips needs to be larger than that used for commons. If you are using 2x6s for the common rafters, use 2x8s for the hips. It's necessary to user larger stock because each hip carries a number of jack rafters and therefore a greater portion of the roof load than any single common rafter. The larger hip will also give full bearing to the jacks, which connect to it at an angle.

Hip-rafter length

The length of hip rafters is calculated on the basis of a 17-in. run, as opposed to the 12-in. run on a common rafter. Both types of rafters have the same vertical rise, from the plates to the ridge, but the hip comes in at a 45° angle, which results in a longer horizontal run. The 17-in. figure is arrived at by calculating the diagonal in a 12-in. square, which is 16.97 in., rounded to 17. The roof pitch can change, but the units of run, 12 for the common and 17 for the hip, remain constant.

An easy way to determine the length of hip rafters is to look under the hip/valley column in a book of rafter tables. On a building with a 4-in-12 pitch and a 20-ft. span, for example, the theoretical hip length will be 14 ft. 6⅜ in. (as shown in the table at right). The theoretical hip length is the distance from the plumb cut at the peak of the roof, center of the ridge, to the plumb cut marking the heel of the seat cut at the building line. Remember that, as with common rafters, the theoretical length does not consider the thickness of the ridge, nor the tail for an overhang. Both of these factors must be calculated before ordering material or cutting rafters to length.

Relationship of Hip to Common Rafters

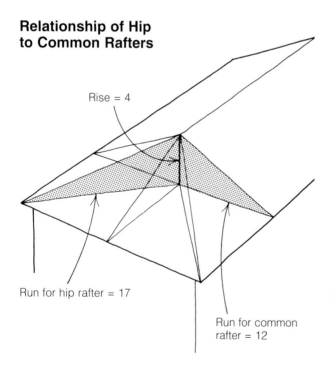

Rise = 4

Run for hip rafter = 17

Run for common rafter = 12

HIP/VALLEY-RAFTER TABLE 4-in-12 pitch (18½° angle)			
Span Ft.	Length Ft. In.	Span In.	Length In.
1	8¾	¼	⅛
2	1. 5⅜	½	⅜
3	2. 2⅛	¾	½
4	2. 10⅞	1	¾
5	3. 7⅝	1¼	⅞
6	4. 4¼	1½	1 ⅛
7	5. 1	1¾	1¼
8	5. 9¾	2	1½
9	6. 6½	2¼	1⅝
10	7. 3⅛	2½	1⅞
11	7. 11⅞	2¾	2
12	8. 8⅝	3	2⅛
13	9. 5⅜	3¼	2⅜
14	10. 2	3½	2½
15	10. 10¾	3¾	2¾
16	11. 7½	4	2⅞
17	12. 4¼	4¼	3 ⅛
18	13. 0⅞	4½	3¼
19	13. 9⅛	4¾	3½
20	14. 6⅜	5	3⅝
21	15. 3⅛	5¼	3⅞
22	15. 11¾	5½	4
23	16. 8½	5¾	4⅛
24	17. 5¼	6	4⅜
25	18. 2	6¼	4½
26	18. 10⅝	6½	4¾
27	19. 7⅜	6¾	4⅞
28	20. 4⅛	7	5 ⅛
29	21. 0⅞	7¼	5¼
30	21. 9½	7½	5½
31	22. 6¼	7¾	5⅝
32	23. 3	8	5¾
33	23. 11¾	8¼	6
34	24. 8⅜	8½	6⅛
35	25. 5⅛	8¾	6⅜
36	26. 1⅞	9	6½
37	26. 10½	9¼	6¾
38	27. 7¼	9½	6⅞
39	28. 4	9¾	7 ⅛
40	29. 0¾	10	7¼
41	29. 9⅜	10¼	7½
42	30. 6⅛	10½	7⅝
43	31. 2⅞	10¾	7¾
44	31. 11⅝	11	8
45	32. 8¼	11¼	8⅛
46	33. 5	11½	8⅜
47	34. 1¾	11¾	8½
48	34. 10½	12	8¾
49	35. 7⅛		
50	36. 3⅞		

The rafters and ridges on most houses are standard 2x material, 1½ in. thick. Like commons, hips need to be shortened one-half the thickness of the ridge, or if they land against a common, one-half the thickness of the common (see p. 138). Unlike commons on a gable roof, which rest against the ridge at a right angle, hips come to the peak at a 45° angle, so they shorten one-half the 45° thickness of the ridge. If you make a 45° mark across the top edge of 2x ridge stock and measure that line, it equals 2⅛ in. (see the drawing below). The hip is calculated to go to the centerline of the ridge. This amount, half the 45° thickness (1¹⁄₁₆ in.), is subtracted from the rafter at a right angle to the ridge plumb cut at the long point of the hip.

Shortening has to be done perfectly for exposed roof rafters. On most residential housing, however, even if every rafter is cut perfectly, some adjustment may have to be made when stacking the rafters because the building may be slightly out of square or the walls slightly out of parallel. So, in practice, especially for lower pitches, roof cutters simply shorten the hip by subtracting 1¹⁄₁₆ in. from the overall length of the rafter. Although it is important to work accurately and do quality work, bear in mind that a hip rafter that is, say, ½ in. long will work fine, is structurally strong, will be covered by roof sheathing and will be seen only by spiders living in the attic.

When buying hip stock, make sure that the boards are long enough to cover the overhang. As with common rafters, the exact length of a hip is easy to determine from a rafter-table book. If the plans call for a 2-ft. overhang, for example, double that figure, which will give you the span, then look at the tables under the hip/valley column at 4 ft. For a 4-in-12 pitch the length from building line to tail end is 2 ft. 10⅞ in. The tails will be cut to length once the roof is stacked so the exposed overhang will be absolutely straight.

If the stock you have is too short for both the rafter and the tail, you can splice two pieces together to get the required length. Lay one piece on top of the

Shortening Hip Rafters

Centerline of ridge

2⅛ in.

X

45°

2x ridge

1¹⁄₁₆ in.

King common rafter

2x hip rafter

Hip rafters are measured to the centerline of the ridge (point X). To fit properly, the hip rafter must be shortened one-half the length of a 45° line across the edge of the ridge (1¹⁄₁₆ in. on a 2x ridge).

Splicing Stock for Long Hips

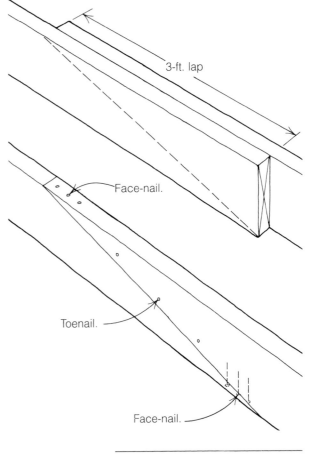

3-ft. lap

Face-nail.

Toenail.

Face-nail.

other, lapping them about 3 ft. Make sure the two pieces lie in a straight line and make a diagonal cut across the lap.

Cut the bottom board by following the saw-kerf mark left from the first cut, then nail the two long pieces together to form one board. Nails can be driven down through the edge and toenailed into the sides. Make this splice as near to the ridge cut as possible so that it can be braced off a wall with a leg or post under it once it is nailed in position.

Dropping the hip

The centerline of the hip rafter is the line where the two slopes on a roof meet. In order for sheathing and fascia to be installed properly, the edges of the hip rafter need to be on the same plane as the jacks and commons. This is accomplished by "dropping" the hip rafter, that is, lowering its elevation by making a deeper seat cut.

The amount that a hip has to be dropped depends upon the thickness of the hip and the pitch of the roof. For a 4-in-12 roof it will be about 3/16 in. One way to determine this amount is to take a piece of rafter stock and mark a level seat-cut line anywhere on the material. From the edge of the stock and along the level line, measure back one-half the thickness of the hip rafter (3/4 in. for 2x stock). Measure at a right angle from this point to the top edge of the material and mark a second level line. The distance between the two level lines is the amount of hip-rafter drop. Lowering the hip allows the sheathing to plane in with the edge of the hip rafter rather than its center. In practice, experienced roof cutters simply drop the hip about 1/4 in. for roof pitches of 4-in-12 through 6-in-12. For a roof with a steeper pitch or a wide hip, the drop will be greater.

Plate layout

The location of hips on most roofs is a known factor—they extend from the corners to the ridge. Surrounding the hips are three common rafters called king commons. The location of the king commons is determined by measuring the span of the building and dividing that figure by two. The result is the run of the rafters. So, for example, if the span is 20 ft., the run will be 10 ft. Measure in from the outside

Dropping the Hip

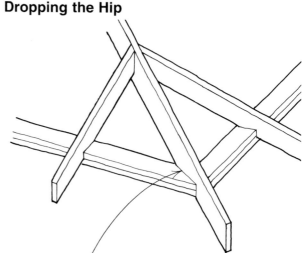

Hip rafters require a deeper seat cut to allow roof sheathing to remain on the same plane as the jacks and commons.

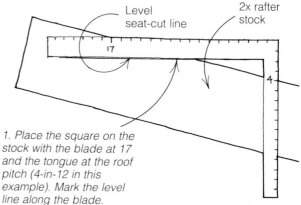

1. Place the square on the stock with the blade at 17 and the tongue at the roof pitch (4-in-12 in this example). Mark the level line along the blade.

2. Measure back one-half the thickness of the hip stock (3/4 in. on a 2x) along the seat-cut line.

3. From this point, square up to the top edge of the rafter stock and mark a second level line. The distance between the two level lines is the amount that the hip rafter needs to drop.

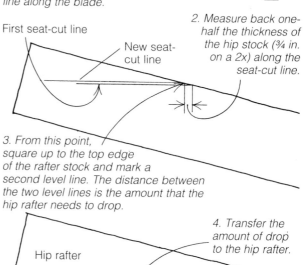

4. Transfer the amount of drop to the hip rafter.

The seat cut is dropped this amount.

Plate Layout

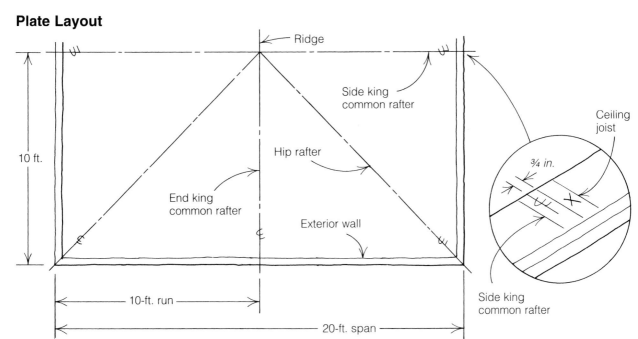

The king common should be centered at a point from the edge that is one-half the span. Mark the layout ¾ in. from this centerline. The joist layout starts at this point.

10 ft. to find the center of the span, which is the center of the end king common, and mark the location on the double top plate. Next, measure down both sides of the span and mark this same distance to the center of the side king commons.

From the center mark, measure over ¾ in., one-half the thickness of 2x rafters, and mark the exact location of the king commons. This is also the point from which ceiling-joist layout is begun on hip roofs (see p. 133). Rafters need to be nailed to joists at the plate line to form the truss that helps hold the building together.

Laying out and cutting hip rafters

Use strong, straight stock for the hip rafters. The roof plan will indicate how many you will need. Place the stock on edge on the rafter horses with crowns up. The quickest way to mark hip rafters is with the template shown in the sidebar on p. 164. Use the template to mark the ridge plumb cut on the end of the stock. Most carpenters like to give the hip rafter a double side cut at the ridge so it will fit snugly between the king common rafters. One way to do this is to set the framing square on the rafter stock at the correct pitch and mark down both sides

of the tongue (so that the marks are 1½ in. apart). When all of the hips have been marked, set your saw at 45° and make the first side cut, as shown in the top drawing on p. 165. Make the second cut in the opposite direction, and the result is a hip with a double ridge side cut.

Flush up the ends of the stock, measure down from the point of the ridge cut the calculated length of the hip rafters, and mark for the heel plumb cut. Shorten the rafters the correct amount, 1¹⁄₁₆ in. for 2x stock, and mark this point on the top edge of the rafters. Next, place the registration mark of the template on the mark for the heel cut on the top edge of the rafter stock and scribe the bird's mouth (see the bottom drawing on p. 165). Remember to drop the hip the required amount, as explained above.

The tail part of the template is cut to the width of the common rafters. Slide the template down the stock and scribe this width on the hip-rafter tails. The bird's mouth is cut just like on a common rafter (see pp. 144-147). Rip the tails down to common-rafter size and move the hips to the corners where they will be nailed on.

Making a hip-rafter template

Begin with a 2-ft. piece of 1x stock or ¾-in. plywood that is the same width as the hip stock. For example, if the hip rafter is 2x8, use a 1x8 for the template. Use a small rafter square or a framing square to scribe the ridge and heel cuts.

If you're using a small rafter square, place the square on the stock and pivot it to the correct pitch number on the "hip-val" line (4 for a 4-in-12 pitch). Then mark along the pivot side for the ridge plumb cut. Move the square down the template about 1 ft. and scribe a second plumb mark for the heel cut, extending the line square across the top of the template to serve as a guide when laying out the rafter stock.

If you're using a framing square, set the tongue at 4 in. and the blade at 17 in. on one edge of the board and mark along the tongue for the plumb cut. Move the square down the template and scribe a second plumb mark for the heel plumb cut of the bird's mouth.

The height above plate (HAP) at the bird's mouth on a hip rafter, before it's dropped, is the same as that on a common. Take the common rafter and measure down from the top edge along the plumb line to the level line of the bird's mouth. HAP for a 2x4 common rafter is about 2½ in.; for a 2x6 common it will be about 4½ in. on a 4-in-12 pitch. Transfer this distance to the heel plumb line of the bird's mouth on the hip template. At this point, lay one edge of the small rafter square (or the tongue of the framing square) along the plumb line and scribe the level seat-cut line perpendicular to the plumb line.

Below the bird's mouth, rip a 6-in. section of the 1x template down 2 in. to the size of the common rafter. The template can be used to mark tails for ripping rather than laying out each tail with a measuring tape and chalkline.

Nail a 1x2 fence on the top edge of the template, making sure to leave the registration mark for the bird's mouth exposed.

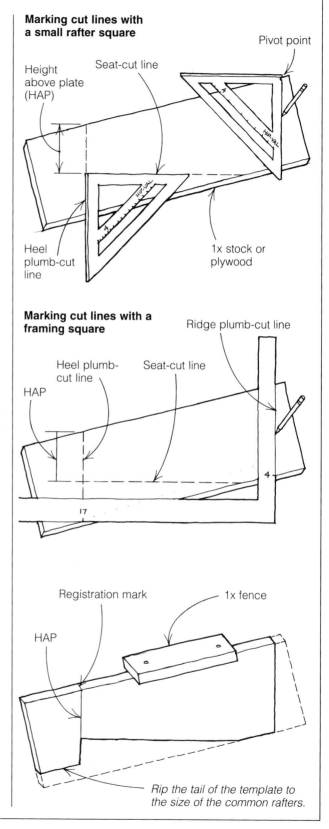

Marking cut lines with a small rafter square

Pivot point

Seat-cut line

Height above plate (HAP)

Heel plumb-cut line

1x stock or plywood

Marking cut lines with a framing square

Ridge plumb-cut line

Heel plumb-cut line

Seat-cut line

HAP

Registration mark

1x fence

HAP

Rip the tail of the template to the size of the common rafters.

Ridge Cuts on a Hip Rafter

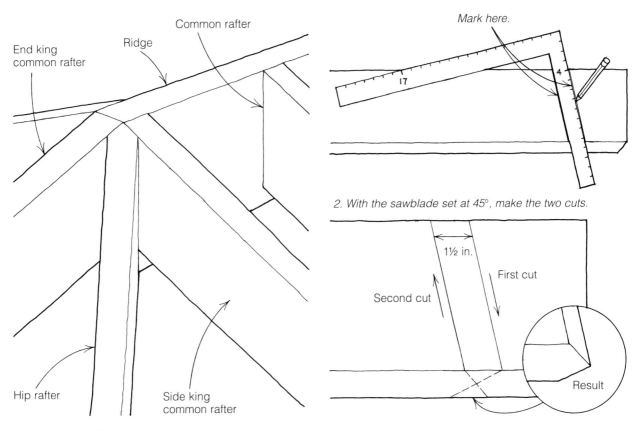

End king
common rafter

Ridge

Common rafter

Hip rafter

Side king
common rafter

1. With the framing square set on the hip-rafter stock at the correct pitch, mark a line down both sides of the tongue.

Mark here.

17

4

2. With the sawblade set at 45°, make the two cuts.

1½ in.

First cut

Second cut

Result

Laying Out Hip Rafters

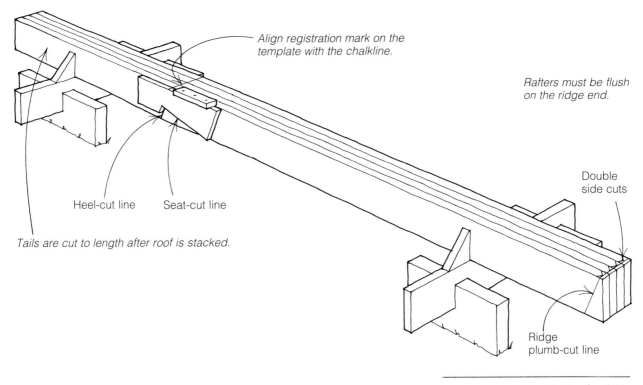

Align registration mark on the template with the chalkline.

Rafters must be flush on the ridge end.

Heel-cut line

Seat-cut line

Double
side cuts

Tails are cut to length after roof is stacked.

Ridge
plumb-cut line

Laying out and cutting hip jack rafters

Hip jack rafters are usually cut from the same size stock as commons. There are several ways to lay out jacks for cutting. One way is to rack them up like gable studs, as shown in the drawing below. Begin with a full-length, unshortened common rafter. Next, you need to calculate the common difference, which is the standard difference in length between jack rafters. This information can be found in your book of rafter tables. For a 4-in-12 pitch roof with rafters spaced 16 in. on center, the common difference is 1 ft. 4⅞ in. (see the table at right). If the rafters are spaced 24 in. on center, the common difference is 2 ft. 1¼ in.

The first set of jack rafters, the longest, will be the common difference shorter than a full-length common rafter. Look at the plans to determine how many sets of jacks you need for each span and place this number alongside a common. For a rectangular building with four hips, you will need stock for four sets of jacks (that is, eight). Next, position the stock for the next set of jacks, using lumber that is about 1 ft. shorter than the previous set. When you've laid out all the jack stock, flush up the tail ends.

Spaced In.	Length Ft. In.		Spaced In.	Length Ft. In.	
1		1	25	2.	2⅜
2		2⅛	26	2.	3⅜
3		3⅛	27	2.	4½
4		4¼	28	2.	5½
5		5¼	29	2.	6⅝
6		6⅜	30	2.	7⅝
7		7⅜	31	2.	8⅝
8		8⅜	32	2.	9¾
9		9½	33	2.	10¾
10		10½	34	2.	11⅞
11		11⅝	35	3.	0⅞
12	1.	0⅝	36	3.	2
13	1.	1¾	37	3.	3
14	1.	2¾	38	3.	4
15	1.	3¾	39	3.	5⅛
16	1.	4⅞	40	3.	6⅛
17	1.	5⅞	41	3.	7¼
18	1.	7	42	3.	8¼
19	1.	8	43	3.	9⅜
20	1.	9⅛	44	3.	10⅜
21	1.	10⅛	45	3.	11⅜
22	1.	11¼	46	4.	0½
23	2.	0¼	47	4.	1½
24	2.	1¼	48	4.	2⅝

JACK-RAFTER TABLE
4-in-12 pitch (18½° angle)

Laying Out Hip Jack Rafters

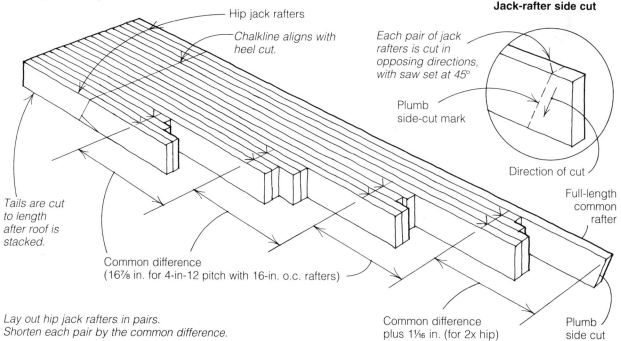

Hip jack rafters

Chalkline aligns with heel cut.

Jack-rafter side cut

Each pair of jack rafters is cut in opposing directions, with saw set at 45°

Plumb side-cut mark

Direction of cut

Full-length common rafter

Tails are cut to length after roof is stacked.

Common difference (16⅞ in. for 4-in-12 pitch with 16-in. o.c. rafters)

Common difference plus 1 1/16 in. (for 2x hip)

Plumb side cut

Lay out hip jack rafters in pairs.
Shorten each pair by the common difference.

Two Jacks from One Cut

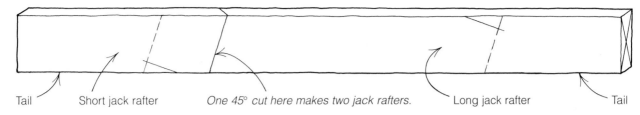

Tail Short jack rafter *One 45° cut here makes two jack rafters.* Long jack rafter Tail

Hip jack rafters need the same length of tail as commons, so measure up from the flushed tail ends and make a mark across the edges of all the stock and mark the heel cut. Rafter tails are usually cut to length after they are stacked, so at this point they need to be left at least 2 in. or 3 in. longer than needed. Next, from the ridge plumb cut of the un-shortened common, measure back the common difference on the first set of jacks. Shorten this set of jack rafters one-half the 45° thickness of the hip (1¹⁄₁₆ in. for a 2x), and mark this length across the top edge of all the jacks in this set. Make a diagonal mark to indicate the direction of the side cut for each pair of jacks that nail to opposing sides of the hip or valley. Measure down the common difference and mark the next set, and so on.

Using the common-rafter template (see the sidebar on pp. 140-141), mark the plumb side cut and the bird's-mouth cut. Because the jack rafters are in pairs, the plumb cut at the hip is laid out on alternating sides of the rafter stock.

Make the cuts and stack the jacks in appropriate piles so they can be carried to the corners and placed up on the ceiling joists. The shorter jacks can be hung from the double top plate by a 16d nail driven into the side cut.

Some experienced framers have developed a shortcut for cutting jacks. They make one plumb ridge cut work for two jacks. This is done by using stock that is long enough for both the longest and shortest jacks. Lay out the longest jack with the bird's mouth on the top edge of the stock and the shortest with the bird's mouth on the bottom edge, as shown in the drawing above.

Stacking a hip roof

A basic hip roof is put together much like a simple gable roof. Run a 16-in. or 24-in. on-center layout, as called for on the plans, on a section of ridge. Bring up the two opposing commons next to the king-common layout at the hip end and nail them to the plate and joists at the building line. Then go to the end of the ridge board and nail in two more commons. Bring the ridge board up between the two sets of common rafters and nail it in place, leaving 16 in. to 24 in. sticking out on the hip end to receive the king commons (see the top drawing on p. 168). There is no need at this point to worry about trying to calculate the exact ridge length—this is more easily done when it's time to nail in the end king common.

Make sure the ridge is level (see p. 150), then cut 2x posts to put under each end to hold it in place. Plumbing the rafters is easy on a gable roof because the end of the building is the reference point. On a hip roof, the commons can be plumbed off a parallel joist and held in position with a sway brace. On most roofs, common rafters tie to parallel joists to form the roof truss. If the commons are directly alongside and in line with the joists, then they are plumb. You can check this with a level, or you can do what a lot of experienced carpenters do—stand up straight and sight the rafter parallel with the joist by eye. This method may not be perfect, but if it looks close enough, nail it.

A long building will need several sections of ridge. These can now be set between commons and held level at the proper height by posts. At each ridge end, bring up an end king common, hold it alongside the ridge, and mark and cut the ridge to length. Nail the end common to the ridge end and the wall plate. It, too, now functions as a sway brace.

Stacking a Hip Roof

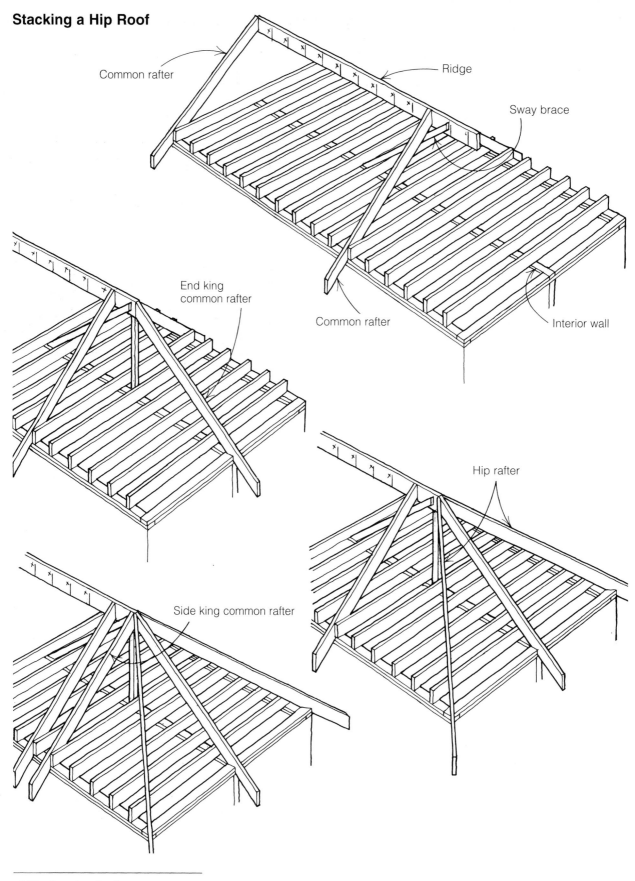

Common rafter

Ridge

Sway brace

Common rafter

Interior wall

End king
common rafter

Side king common rafter

Hip rafter

With the hip rafter in place, the jack rafters can be nailed on, beginning with the longest one.

Hip rafters need to be held straight as the jack rafters and frieze blocks are nailed in place.

Next, pull up a hip rafter and nail it at the wall-plate line directly over the outside corner with two 16d toenails on one side and one on the other. At the ridge, the side cut will lie flat against the end common. Drive three 16ds through the hip into the common. Pull up the opposing hip and nail it in place. Pull up the two side king commons and nail them at the plate and into the ridge and hip with two more nails. Do the same with the other hips, and the bare bones of the roof are in place.

Take time to support the hips by nailing 2x posts from an intersecting load-bearing wall up under them, just as with purlins on a gable roof (see pp. 152-153). It is especially important to support long or spliced rafters. The supports will keep them aligned (with no crown or sag) until the jacks are nailed in place.

Before nailing in the jacks, sight down the hip rafter and make sure it is straight from the ridge to the plate. If it isn't, put a temporary brace in the middle to hold it straight until the jacks are nailed in place. Then, beginning with the longest jack, nail it in on layout at the plate along with a frieze block (see p. 147). Nail each jack to the hip with three 16d nails, taking care not to bow the hip rafter from side to side. Once the opposing jack is nailed in, the hip will be permanently held in place.

Frieze blocks at the corners have to have a side cut to fit snug against the hips or valleys. Once all the jacks and commons are nailed in, the overhangs can be measured, marked and trimmed to length, and the fascia can be nailed on if the plans call for it. The length of the hip overhang is determined simply by extending the chalkline on the commons all the way across the hip. When marking tails to length, be sure to use the common-rafter template on the commons and the hip-valley template on the hips. Once the required purlins, rafter ties and fascia are nailed in place, the roof is ready to sheathe.

INTERSECTING ROOFS

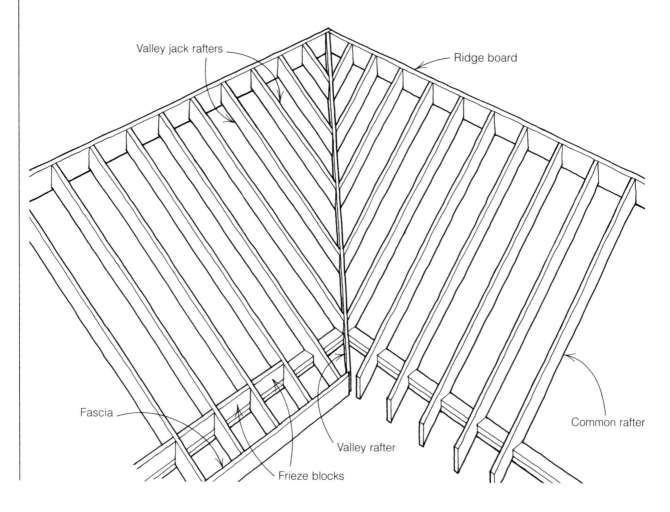

Valley jack rafters

Ridge board

Fascia

Common rafter

Valley rafter

Frieze blocks

Whenever two sections of a building come together at a right angle, the roof planes intersect and create a valley. Traditional valleys are constructed with valley rafters. The space between the ridge and the valley rafter is filled with jack rafters. The blind-valley system, also called a "California roof" (see pp. 173-176), differs from traditionally framed valleys in that it does not require valley rafters.

Traditional valleys are much like inverted hips. Both hip and valley rafters are the same length for the same span, they are cut from the same-size lumber using the same template, they each have a constant run of 17 in., and they shorten the same way. Valleys differ from hips in that they run from the ridge to inside corners rather than outside corners, they don't require dropping (lowering), and at the point where the valley meets the ridge, the plumb cut is usually given a single side cut.

Valley jack rafters

Valley jack rafters differ from hip jack rafters in that they have no tail, because they run from the valley rafter to the ridge, and are shortened both one-half the 45° thickness of the valley (1¹⁄₁₆ in. for a 2x) and one-half the thickness of the ridge (¾ in. for a 2x). (For more on shortening rafters, see pgs. 138 and 161.) At the valley, a valley jack gets a side cut like a hip jack and at the ridge, a plumb cut like a common rafter.

Intersecting Roofs with Equal Spans

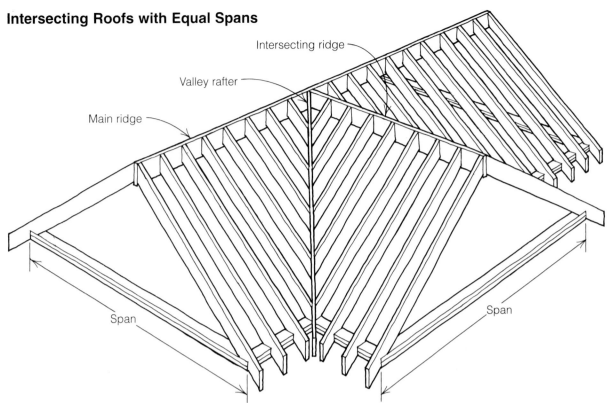

Intersecting ridge

Valley rafter

Main ridge

Span

Span

If the two roof sections are the same pitch and have the same span, the ridges will be at the same height.

When parallel hip and valley rafters are close together, the space between them is framed in with hip-valley cripple jack rafters. All of these rafters are the same size. An easy way to determine their length is take their run, then double the figure so you have the span, and look in a rafter-table book under the common-rafter column (see the table on p. 139). If, for example, the run is 3 ft., the rafter span is 6 ft. and, for a 4-in-12 pitch, the length is 3 ft. 2 in. With 2x stock, the valley jacks need to be shortened 2⅛ in., that is, one-half the 45° thickness of the valley rafter and one-half the 45° thickness of the hip rafter. Both ends have a plumb cut with a single side cut to fit against the valley or the hip.

Equal spans

When the intersecting building sections are the same width they have equal spans, and the ridge height for both sections will be the same (see the drawing above). Stack the main section just as you would a normal gable or hip roof. To determine the length of the valley rafters, look under the hip/valley column in a rafter-table book (see p. 160). Then, to find the location for the intersecting ridge, pull up the valley rafters and mark where they land against the main ridge. The intersecting ridge nails between the valley rafters to the main ridge and is held level by posts (see p. 167). Nail in all the commons and jacks, and the roof is basically framed.

Unequal spans

When the intersecting roofs cover building sections with unequal spans, connecting the two is more complicated. A common way to frame this type of roof is to frame both roofs as separate entities and connect them with a broken or continuation hip (see the top drawing on p. 172).

To determine the length of the broken hip, subtract the hip-rafter length of the minor span from that of the major span. For example, on a 4-in-12 roof, a 20-ft. major span needs a hip of 14 ft. 6⅜ in., and a 14-ft. minor span needs a 10-ft. 2-in. hip. The difference in length between the two is 4 ft. 4⅜ in., which is the theoretical length of the broken hip. It

Intersecting Roofs with Unequal Spans

If the spans on the two sections are different, the ridges will be at different heights.

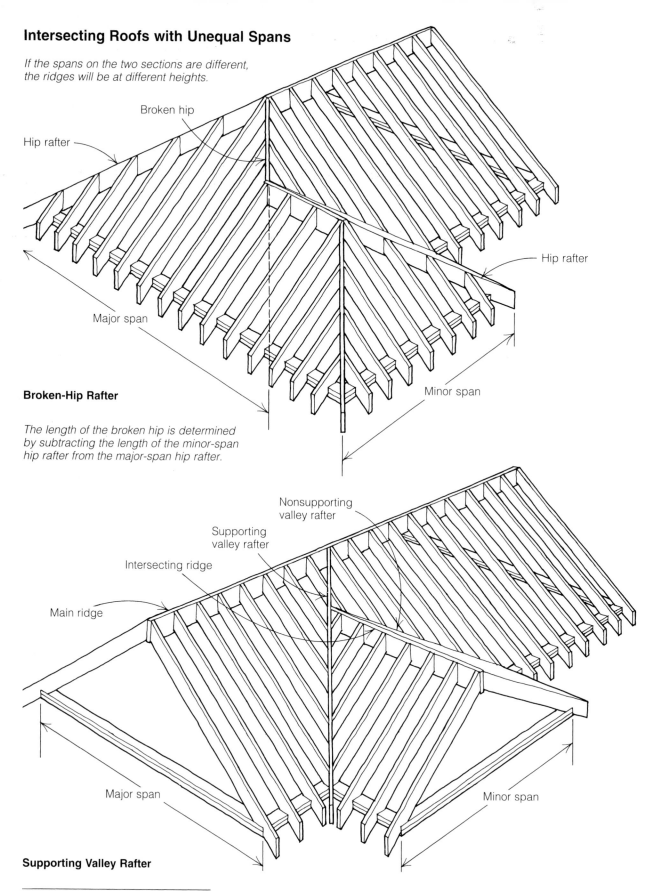

Broken hip

Hip rafter

Hip rafter

Major span

Minor span

Broken-Hip Rafter

The length of the broken hip is determined by subtracting the length of the minor-span hip rafter from the major-span hip rafter.

Nonsupporting valley rafter

Supporting valley rafter

Intersecting ridge

Main ridge

Major span

Minor span

Supporting Valley Rafter

has to be shortened 2⅛ in. because it goes from ridge to ridge at a 45° angle. Finally, the broken hip requires a plumb cut at both ends with side cuts.

Another way to cut and stack an unequal-span roof is to use a supporting valley intersected by a nonsupporting valley (see the bottom drawing on the facing page). The supporting valley rafter runs from an inside corner to the main or highest ridge, and its length is calculated from the main span and is shortened like a regular valley. Try to support it in the middle by running a post from the valley down to a bearing wall.

The nonsupporting valley runs from the other inside corner of the building and hits the supporting valley at a right angle, fitting against it with a square plumb cut. It is shortened like a common rafter, one-half the thickness of the supporting valley. The length of this valley is calculated from the minor span.

A valley rafter sits square in an inside corner. Unless it has angled cuts at the bird's mouth, it can't be moved into the corner. Most roof cutters find no need to make these angled cuts unless they're working with beam stock with exposed rafters.

Valley Jack Rafters

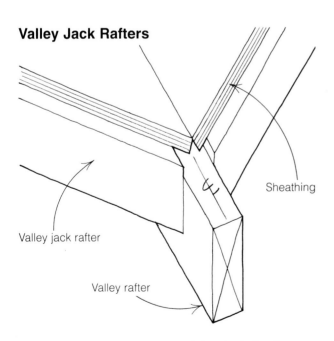

Valley jack rafter

Sheathing

Valley rafter

Valleys do not have to be dropped, but the jack rafters need to be held high enough so that sheathing meets at the centerline of the valley rafter.

Unlike a hip rafter, a valley rafter does not need to be dropped. But the jack rafters on a valley should be nailed in a little high, the same amount a hip is dropped, to allow the sheathing to plane in at the center of the valley rafter (see the drawing below). An easy way to determine this amount is to hold a short straightedge on top of a valley jack nailed in place to the ridge and allow it to plane in with the center of the valley rafter. Hold all valley jacks up the same amount. Mark and cut valley-rafter tails to length in the overhang just like a hip.

Blind ("California") valleys

The blind-valley system is a way for one roof to intersect another without the need for valley rafters. It is the preferred method whenever you add a new wing to an old house. Rather than tear off the old roof and cut in valleys, it's much easier to let the new roof intersect the old with a blind valley. Even on a new house it can be the fastest method, as, for example, when false dormers are framed on roof slopes. Use a blind valley whenever the rafters of the main roof will be sheathed on the underside to form a cathedral ceiling.

To construct a blind-valley roof, it is first necessary to stack the common rafters in the main roof and sheathe at least enough so that the blind valleys fall on the sheathing (see the drawing on p. 175). Some codes require that the entire main roof be sheathed, even under the secondary roof, in order to maintain maximum shear value.

Once the main roof is stacked and sheathed, continue by stacking the commons on the secondary roof and extending its ridge over to the sheathed area of the main roof (as shown in the top photo on p. 174). When the extension ridge touches the main roof, mark it to length. Hold another piece of ridge stock alongside the extension ridge on the main roof and scribe the pitch on the extension. Make the cuts and nail the extension ridge in place on top of the sheathing, making sure to sight it both level and straight.

Now snap a chalkline for the valley by holding the top of the line at the point of the extended ridge and pulling the other end down to the inside corner. This point is where the common rafters for both roofs plane together. Along this line, additional sup-

Top: The blind-valley ridge can be marked to length by extending a second piece of ridge to the main roof. ***Above left:*** Lay another piece of ridge stock alongside the extension on the main roof and scribe the pitch of the roof on the extension. ***Above right:*** Cut along the pitchline and the extension can be nailed in place.

Blind ('California') Valley

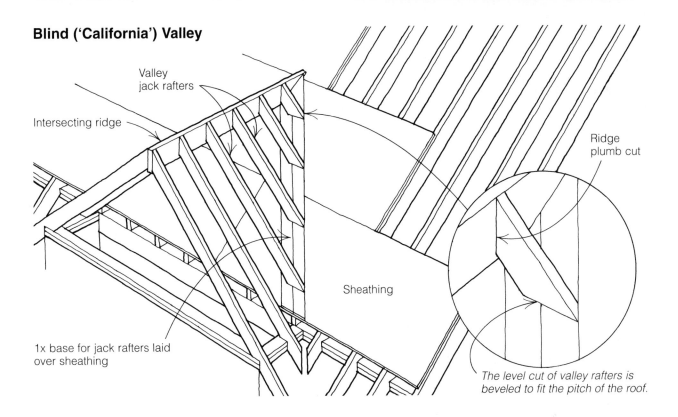

Valley jack rafters

Intersecting ridge

Ridge plumb cut

Sheathing

1x base for jack rafters laid over sheathing

The level cut of valley rafters is beveled to fit the pitch of the roof.

port needs to be added as a base for the bottom ends of the valley jack rafters. Jacks should not be nailed directly to the roof sheathing, so nail two 1x6s, or pieces of plywood about that width, alongside the valley line, holding them back far enough so that the top edges of the rafters will plane in with the line and the rest of the roof. The steeper the pitch, the closer to the line the boards will be. To figure the exact distance, pull a line from the ridge of the intersecting roof to the chalkline in the valley and push your 1x or plywood up to it.

The space between the last common on the ridge and the end of the valley is now filled in with jack rafters, which are laid out in pairs set on edge and shortened the common difference (see pp. 166-167). Like regular valley jack rafters, these have a plumb cut at the ridge. At the bottom end they have a level cut on a bevel to fit the pitch of the roof. The bevel at the bottom tips either to the right or to the left, depending on which side of the ridge the jack is to be nailed. Make a diagonal mark to indicate the direction of the bevel cut for each pair of jacks.

The quickest way to mark blind-valley jack rafters is with the template shown in the sidebar on p. 176.

With the jack rafters still on edge, use the template to mark the ridge cut on the end of the stock. For the level cut, place the template so that its point is even with, and on the same side as, the point where the length mark and the diagonal mark come together on the jack. Each member of each set of jacks, therefore, gets the level cut marked on opposite sides.

The ridge is cut like a common rafter. To cut the bevel for the level cut, set the saw to the correct angle for the roof pitch (18½° for a 4-in-12 pitch) and follow the marks. Make sure the sawblade tips in the same direction as the diagonal mark: one to the left, one to the right.

Now nail the jacks in place on the ridge following the ridge layout (see the photo on p. 176). Nail the bottom end, the level cut, into the 1x base laid over the sheathing. Toenail rather than nail straight down to save your sawblade when cutting sheathing.

If the roofs intersect at the same plate height, a dummy valley tail needs to be cut into the overhang to give the sheathing something to break on. The 1x base that forms the valley on the main roof can't be

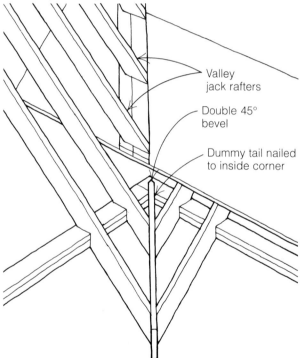

The blind-valley jack rafters are nailed from the ridge to the 1x base, which is secured to the main roof.

Dummy Valley-Rafter Tails

Valley
jack rafters

Double 45°
bevel

Dummy tail nailed
to inside corner

Common rafters are cut at 45° angle to allow dummy tail to fit between rafters.

SITE-BUILT TOOLS

Making a blind valley jack-rafter template
The blind valley jack-rafter template is much like the templates for common rafters (see pp. 140-141) and hip rafters (p. 164). On one end of a short 1x, the same width as the rafter stock, scribe a ridge plumb cut. On the other end, about 12 in. away, scribe a level cut perpendicular to the plumb cut. Nail a 1x2 fence on the top edge to make it easy to place the template on the jacks and mark the cuts.

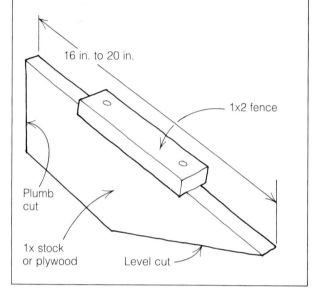

16 in. to 20 in.

1x2 fence

Plumb
cut

1x stock
or plywood

Level cut

With the valley stock in place, scribe the tails of the common rafters to length by marking along both sides of the valley stock. Place the common-rafter template on these marks and scribe the plumb-cut line. Then make the cuts with your saw set at 45°. Next, cut a double 45° plumb cut on one end of the dummy tail so that it will fit into the corner, drop it down between the rafters and nail it in place. Let the end of the valley run wild and mark all the tails to length with a chalkline. Mark the valley and common rafters with their respective templates and make the tail cuts.

If one roof intersects another at different plate heights, you may need to fit a bird stop, a scrap of plywood or 2x, behind a rafter and between the building line and the fascia. This stop will keep birds from nesting under open eaves and simplify shingling.

run into the overhang. The simplest way to cut the dummy tail is to lay a piece of rafter stock on edge right in the valley, letting it extend out into the overhang and over the common-rafter tails.

TRUSS ROOFS

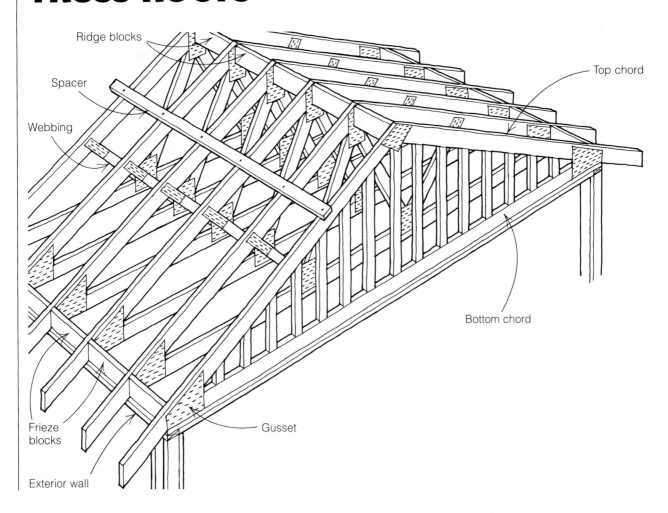

Ridge blocks

Spacer

Webbing

Top chord

Bottom chord

Frieze blocks

Exterior wall

Gusset

Roofs built with factory-built trusses have become popular in residential housing construction in many parts of the country. Trusses can be erected quickly and generally require less labor and material than roofs built on the job site. Trusses can be engineered to span larger distances than conventional stick-framed roofs, allowing more flexibility in room sizes. One disadvantage of most roof trusses is that, because of their design, they sharply reduce the usefulness of an attic for storing family heirlooms or later conversion to living space.

Commonly used trusses have top and bottom chords with webbing in between, and are usually made of 2x4s. The top chords are the sloping rafters that form the roof line. The bottom chords are the horizontal joists that form the ceiling. The webbing helps connect the chords and transfer the load from one member to another and to bearing points designed by an engineer. The chords and webbing are tied together with metal (sometimes plywood) connector plates called gussets.

There are many different types of trusses. Usually they are built to bear on the two exterior walls, sometimes with additional support from an interior bearing wall. Some common roof-truss designs are shown in the drawing on p. 178.

Common Truss Designs

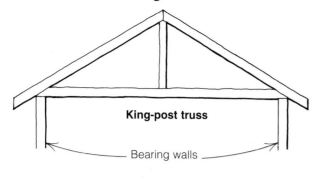

King-post truss

Bearing walls

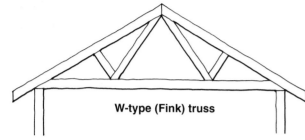

W-type (Fink) truss

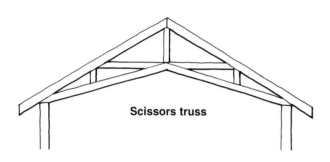

Scissors truss

When building a conventional roof you can often order your wood one day and have it delivered the next. When using trusses it is important not to wait until the walls are framed to place your order. Many manufacturers require several weeks lead time to build and ship trusses. When you place your order, meet with the manufacturer's representative and make sure that your list is complete and all the measurements are accurate. If the roof has any skylights or other large openings in it, extra trusses will be needed to support the roof in these areas. Frequently the truss company also supplies the blocks needed at the plate and ridge line. Once the order is firm, a delivery date can be set.

Preparation

All the walls in the building must be plumbed, lined and well braced. Cut frieze blocks or ridge blocks if they aren't supplied by the truss company. Most trusses are installed 24 in. on center, and the standard block for this spacing is 22½ in. long. Often the blocks will hit the gussets both at the plate line and the ridge and need to be shortened accordingly. Get it right and keep the truss units properly spaced. A mistake here means that sheathing and drywall won't land in the center of a rafter or joist.

On an average house, two carpenters can easily set, block and brace the roof trusses. But on a house requiring extra-long trusses or with a steep roof pitch, a third person may be needed in the middle to help move the truss to an upright position. If so, a catwalk or scaffold may need to be constructed on the floor below so that the third person has something to stand on.

Loading and scattering trusses

Trusses are delivered in bundles; one bundle for each wing of the house is a standard procedure. This way they can be unloaded near to where they will be installed. Some companies have a small crane to unload the trusses right onto the wall plates (which, because of the load, is one reason the walls need to be plumbed and well braced). If no crane is available, the trusses can be set in place with a forklift, or even lifted into position one at a time by hand.

It's usually most efficient to have the trusses placed flat across the walls at the end of the house opposite to where the installation will begin. The trusses should be loaded with their ridge hanging over the outside of the building as far as possible (see the bottom photo on the facing page). If a rake wall prevents the trusses from being loaded in this manner, the truss bundles can be set upright on the walls. Before cutting the metal bands that hold the bundle together, wrap a rope or wire around the top chords near the ridge, so that once the bands are cut the bundle can expand and all the trusses will be held upright. Otherwise they can fly off in unexpected directions, perhaps on top of you.

Trusses are delivered to the job site in large, tightly banded bundles. Take care when cutting the bands—they can be under a lot of pressure.

Remember that trusses are engineered units and that any cutting or drilling in them may damage their structural integrity. Never modify a truss without checking with the manufacturer or engineer to see if it is permissible. An exception to this rule is that the gable-end trusses, which will normally have gable studs in them rather than webbing, can have their top chords notched for 2x lookouts to carry the barge rafters (see pp. 155-156).

Installing trusses

Before moving the first truss into position, drive a nail into each frieze block and hang the blocks from nails driven every 2 ft. along the top plate on the outside walls. The plans or code may call for some blocks to have screened vents in them to allow air into the attic. These vents are generally placed 2 ft. from each end and every 6 ft. in between. Notch the gable-end truss for lookouts, if required by the plans.

Next, nail a long, straight 2x on edge against the side of the wall frame at the far end of the building. The 2x should be installed near the center of the ridge and extend above the top chord. When the gable-end truss is raised, it can be nailed to this 2x to hold it plumb. Now you can move the first truss

Trusses will usually be much easier to install if they are loaded right on top of the walls, with the ridge hanging out as far as possible.

Before installing the gable-end truss, nail a long 2x4 against the wall to help hold the truss plumb.

Installing the Gable-End Truss

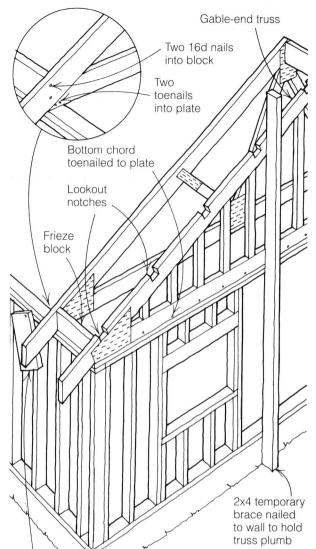

Gable-end truss

Two 16d nails into block

Two toenails into plate

Bottom chord toenailed to plate

Lookout notches

Frieze block

2x4 temporary brace nailed to wall to hold truss plumb

Hang frieze blocks from top plate so they're at hand when installing trusses.

to the end of the building and set it upright in position, flush with the outside of the end wall.

Toenail the truss to the top plates with 16d nails every 16 in. on center. Then fasten it temporarily to the vertical 2x on the end wall. This should be sufficient temporary bracing unless the wind is blowing or the trusses are tall. Often the gable studs are held in place by staples. If so, it's a good idea to nail a 2x4 across them temporarily 4 ft. above the plate line. This 2x acts as a guard rail to keep workers from leaning against a gable stud and falling.

Next, nail frieze blocks into the first truss at both outside plate lines. Bring the second truss into position, toenail a ridge block at the peak, set the truss upright, drive two 16d nails through it into the frieze blocks and sink two 16d toenails down through the bottom chord into the top plates. At this point the two trusses will be fairly stationary. Continue rolling and nailing on trusses. Measure across the top chords at the plate line to make sure the trusses are spaced properly. Trusses can be further braced at the ridge with several lengths of 1x4s,

marked with the on-center layout. These temporary braces can be nailed across the top chords with 8d nails as each truss is rolled into position. The 1x4s will ensure that the rafters run parallel and have the proper spacing until the ridge blocks are fully nailed in place.

If the building has a wing, stack the adjoining roof section with trusses, just as you did on the main building. The area between where the two roof spans meet can be filled in with blind valleys, as described on pp. 173-176. The truss company usually supplies this fill; if not, you can cut it on the job.

Trusses cannot be cut and headed off without checking with the manufacturer or engineer. If the plans call for a 30-in. by 30-in. attic access in the ceiling and the trusses are running 24 in. on center, you need to change the access to, say, 22½ in. by 40 in., or you can frame a 30-in. by 30-in. entry on the outside of the building in a gable end. If a larger opening is required, such as for a skylight, extra trusses may be needed on each side of the hole.

After all the trusses have been rolled and nailed in place, turn and backnail one more 16d toenail through the joist chord down into the top plate. Also add two more 16ds into the other end of each ridge block.

Truss bracing

The proper design for bracing for any truss system should be provided by the manufacturer or engineer, or be spelled out in your code. The bracing will vary depending upon many factors, such as the style of the trusses, the roof span, the type of roof covering and the possibility of earthquakes, snow loads or high winds. Don't try to guess your way through truss bracing.

Sway braces help to hold the roof plumb. Truss rafters are plumbed and braced just like common rafters on a gable roof (see pp. 150-151). On trusses, a 2x sway is run from a ridge block down to the double top plate at a 45° angle. Another kind of brace, a catwalk, ties the bottoms of the trusses together. This brace is frequently made from long 1x stock nailed across the bottom chords or webbing holding the trusses at the proper spacing.

Securing trusses to interior walls

Trusses that rest on an interior load-bearing wall can be nailed to the wall with a 16d toenail on each side. Be sure to maintain proper spacing. The plans may also call for a row of blocks at this point. Toenail the trusses to all other cross walls in the same way.

Trusses that are engineered to rest on outside walls without any interior support have to be treated differently, because they normally have a camber, or crown, to them, which will straighten out under expected loads, such as roofing or snow. These trusses

Bracing for Trusses

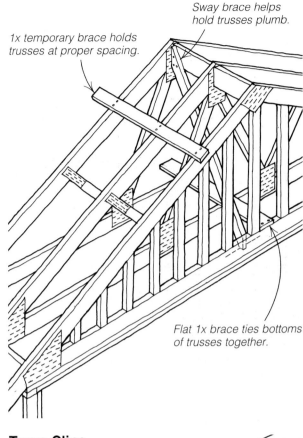

Sway brace helps hold trusses plumb.

1x temporary brace holds trusses at proper spacing.

Flat 1x brace ties bottoms of trusses together.

Truss Clips

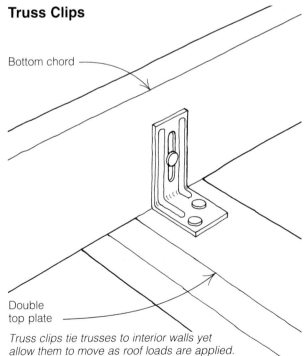

Bottom chord

Double top plate

Truss clips tie trusses to interior walls yet allow them to move as roof loads are applied.

need to be left somewhat free of interior walls so that they can move up and down. They are attached to cross walls with truss clips, which are small connectors that allow the truss to move as roof loads change (see the bottom drawing on p. 181). Other framing anchors, such as hurricane ties, may also be required to secure the trusses to the outside plates or walls. Check the plans and code.

Drywall backing

Drywall backing for parallel walls for trusses that bear on interior walls is installed just as on regular ceiling joists (see p. 54). Backing for free-spanning trusses must be secured to the bottom chords rather than to the top plates, so that the backing will move with the chords once the roof is loaded. This way it will be at the same level as the bottom of the chords.

Most framers nail short 2x4s across the tops of the bottom chords that fall on each side of a parallel partition. The 2x4s are nailed at each end of the wall and every 6 ft. to 7 ft. in between. Then a long 2x the same size as the chord is hung from the short 2x4s parallel to the wall. One long 2x gets nailed on each side of the wall for drywall backing (see the top drawing at right).

Another technique is to use "ladder" backing, with blocks nailed flat between the chords, flush with the bottom edge, at each end and every 24 in. along the full length of an interior partition (see the bottom drawing at right). This backing runs at a right angle over the partition. The blocks can be tied to the partition with truss clips.

Once the interior bracing and backing are in place, the exterior of the roof can be completed like a standard gable roof.

Drywall Backing

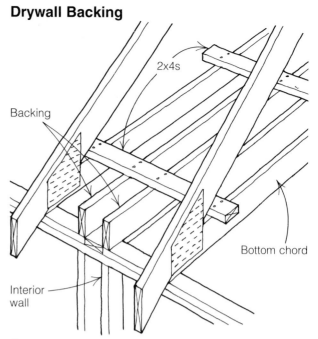

Drywall backing for interior walls can be hung from short 2x4s nailed across bottom chords.

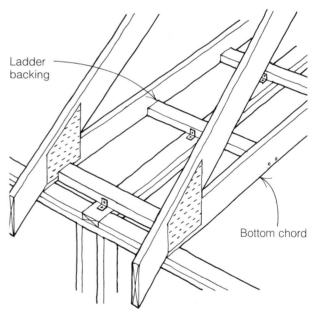

Ceiling drywall can be attached to 'ladder' backing, which is created by nailing 2x4s flush with the bottom edge of the bottom chords.

SHEATHING ROOFS

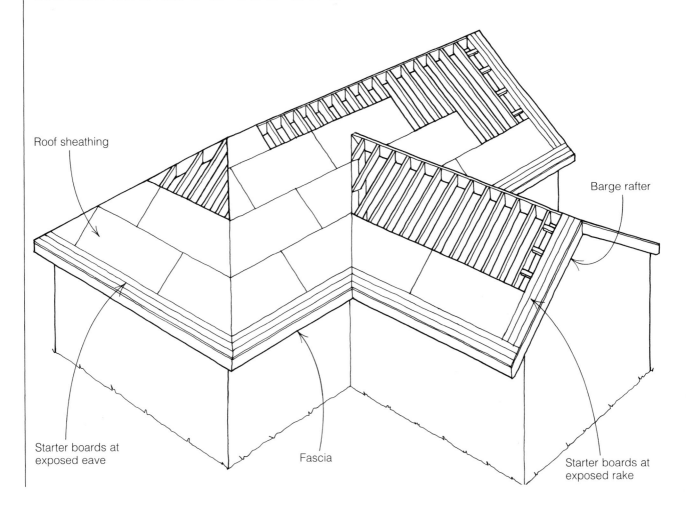

Roof sheathing

Barge rafter

Starter boards at
exposed eave

Fascia

Starter boards at
exposed rake

The type of sheathing that goes on a roof depends to a certain extent on whether or not the eaves are exposed. Most roofs are sheathed with ½-in. or ⅝-in. plywood or oriented strand board (OSB), and the process is much the same as sheathing a floor (see pp. 55-62). If the eaves are going to be closed in with a soffit, the whole roof will receive this sheathing. However, if the rafter tails are going to be exposed, as is often the case in warmer climates, then starter boards may be called for. Since the sheathing on exposed eaves is visible from below, builders like to use a more attractive material than common sheathing panels, such as a better grade of plywood or 1x6 or 1x8 boards (shiplap or tongue and groove).

Installing starter boards

Shiplap boards and finish-grade plywood are easier to install than tongue-and-groove boards. The exposed sheathing can be held flush with the fascia or it can extend over it ¾ in. or so. The most common method is to hold it flush with the fascia and the barge rafter, and then to cover these joints with eave flashing before shingling.

Begin by scattering the starter-board material around the building. If you're using plywood, install it just like floor sheathing, making sure to put the finish side down. When using 1x boards, there is generally no need to snap a control line as long as they are held flush with the fascia. With shiplap, lay the first board so that the lap on the second will fall

Sheathing on a roof

When sheathing a roof, follow these basic safety rules:

• Be careful when working near the edge of the building.

• Walk with care on panels that have sawdust on them.

• When using OSB with a slick side, put the slick side down.

• Don't carry sheathing panels in a strong wind.

• On windy days, nail each sheet with at least six or eight nails to hold it secure until final nailing.

• Secure your tools so they won't slide off the roof.

• Never throw scrap wood off the roof without checking to see that no one is below.

Sheathing an Exposed Eave

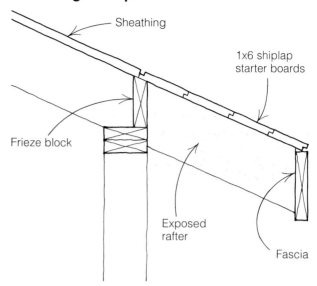

Run the starter boards on an eave up to the frieze block.

Lap Now, Cut Later

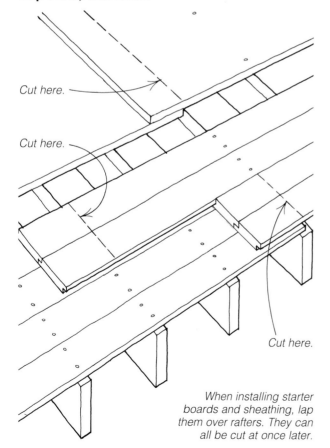

When installing starter boards and sheathing, lap them over rafters. They can all be cut at once later.

on top, and not have to tuck under, as shown in the drawing above. With tongue-and-groove boards, place the tongue facing up the roof.

Sheathe the entire overhang, from the fascia up to the frieze blocks, letting it run wild over the gable end. You want to break all sheathing ends on a rafter, but rather than cut them to length now just let one end break on the center of the rafter and lap the other end on top of it (see the drawing at left). The entire roof can be sheathed this way with panels or shiplap boards, after which you can go back and mark and cut the joints. Tongue-and-groove boards have to be cut as they are laid. With boards, don't allow more than two consecutive joints on one rafter. Boards can be nailed on with two 8d nails into each rafter for 1x6s and three for 1x8s.

There are two ways to sheathe an exposed overhang on the gable end (see the drawing on the facing page). One way is to continue with shorter pieces of starter board beginning at the first inboard rafter, crossing the end rafter and nailing to the top of the barge. The other way involves running the boards up the rake, parallel to the barge, over the lookouts. This method saves time and material. Nail the first board flush with the outside of the barge rafter, and continue until the gable-end eave is covered, breaking the joints on the lookouts. The last board will

Two Ways to Sheathe Exposed Rakes

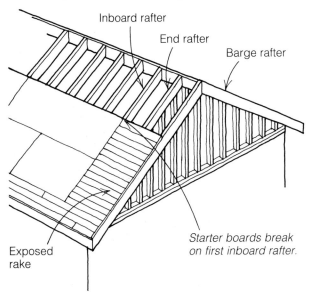

Inboard rafter

End rafter

Barge rafter

Starter boards break on first inboard rafter.

Exposed rake

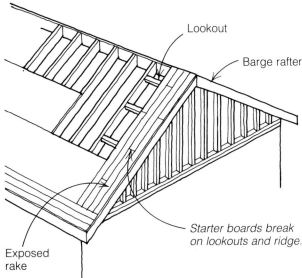

Lookout

Barge rafter

Starter boards break on lookouts and ridge.

Exposed rake

have to be ripped down the center of the rake or gable-end rafter so that the sheathing covering the rest of the roof will have bearing on this rafter.

Panel sheathing

The easiest way to get the rest of the sheathing material onto the roof is with a forklift. If one is not available, you can build a simple platform on which sheathing panels can be stacked so that they can be reached from the roof (see the sidebar at right).

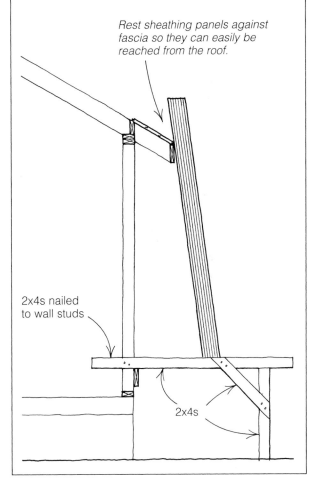

SITE-BUILT TOOLS

Building a platform for roof sheathing

For a one-story house, nail two studs on edge into two wall studs less than 4 ft. apart, extending out from the building about 5 ft. and about 12 in. to 18 in. above floor level. Nail on legs from the ground to support the studs. You now have a platform on which to stack sheathing panels. Rest the panels against the rafter tails or fascia boards to make it easy to grab one and pull it up.

This method can work on some two-story houses if you are able to build the platform on a balcony, making it possible to move the sheathing up in stages.

Rest sheathing panels against fascia so they can easily be reached from the roof.

2x4s nailed to wall studs

2x4s

Begin by snapping a control line across the roof 48¼ in. from the edge of the fascia or starter board. If the sheathing needs to overhang the fascia, make sure to start your measurement from that point. Lay the first course all the way through, breaking joints only at the centers of the rafters and leaving ⅛-in. expansion joints where required.

If a sheet doesn't break over a rafter, lap it back over the one already laid. You can go back and cut off the overlaps once the entire roof is sheathed (see the bottom drawing on p. 184). Stagger the joints just as you did on the floor.

Sheathing panels that end in a valley need to be cut to fit exactly to the center of the valley rafter. An easy way to do this is to shove a full sheet into the valley until one corner touches the valley rafter, and measure the distance from the upper corner to the valley rafter, as shown in the photo below. Transfer the measurement to the bottom edge. Snap a chalkline from the top corner to this mark, cut and install. The cut-off piece can be used as a template to mark other sheets.

Lay the first course of sheathing along a control line, making sure that the joints break over rafters.

Sheathing a valley: *1. Lay the sheathing panel in place and measure the distance from the upper corner to the valley rafter.*

2. Transfer the measurement to the bottom edge of the panel.

3. Connect this point to the upper corner with a chalkline.

4. Cut along the chalkline and the panel should fit the valley perfectly.

Skip Sheathing

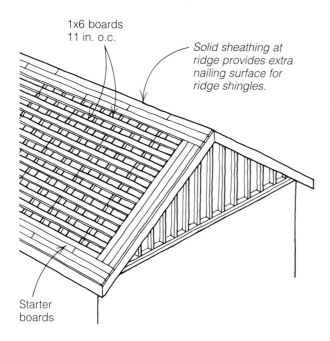

1x6 boards
11 in. o.c.

Solid sheathing at
ridge provides extra
nailing surface for
ridge shingles.

Starter
boards

*Cedar shingles are applied over skip sheathing, which
allows them to 'breathe.'*

Panel Edge Clips

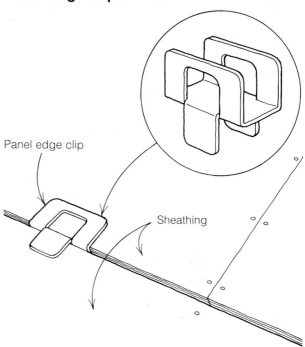

Panel edge clip

Sheathing

*Panel edge clips support sheathing panels
when rafters are spaced 24 in. o.c.*

When sheathing a roof so steep that you can't stand on it, the first row can be installed by standing on the ceiling joists. Then nail long 2x4 cleats through the sheathing and into the rafters to hold you as you work up the roof.

When rafters are spaced 24 in. on center, the sheathing panels may need to be supported between the rafters with a panel edge clip (depending on the code and the type and thickness of the sheathing). The clip fits over the edges of the two panels and links them together so that any load is carried by both sheets.

Roof sheathing is nailed on much like floor sheathing (see pp. 61-62), except that some codes require nails into the ridge and into the frieze blocks between the rafters at the plate line. Always check the code or plans for proper nailing schedules.

Skip sheathing

Skip, or spaced, sheathing is used on roofs that will be covered with cedar shingles, which need to breathe. One common form of skip sheathing uses 1x6 boards spaced on themselves (that is, 5½-in. boards alternating with 5½-in. spaces). That way, you can use a 1x6 as a spacer between courses. At the ridge, sheathe the last three courses solid to give extra nailing to the ridge shingles.

BUILDING STAIRS

Straight-Flight Stairs

Stairs with a Landing

Stairs with a Winder

6

STRAIGHT-FLIGHT STAIRS

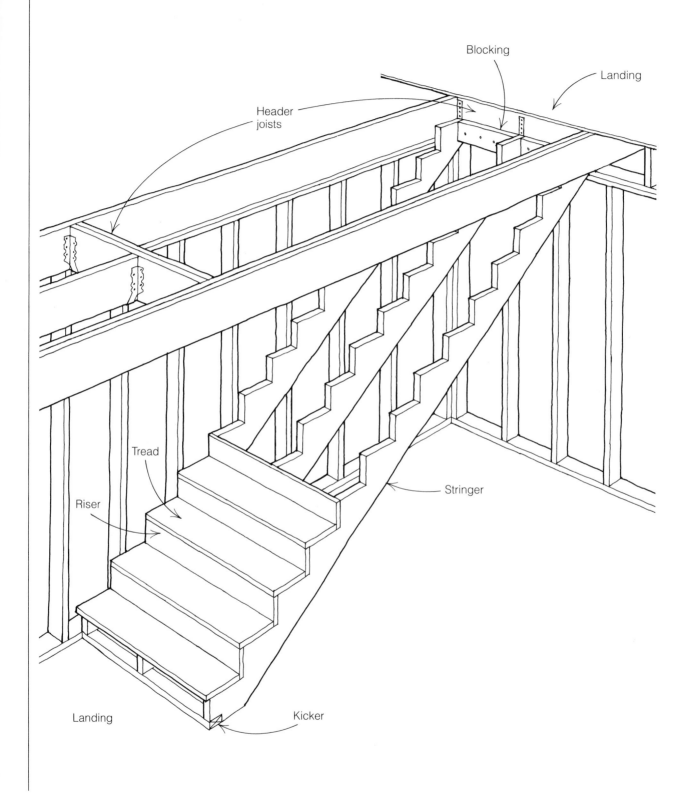

Header joists

Blocking

Landing

Tread

Riser

Stringer

Landing

Kicker

Whether it's with notches cut in the side of a cliff, knots tied in a rope, ladders tied together with rawhide strips, or beautifully crafted stairs in palaces, people have always found ways to get from one level to another. But no matter how they do it, they've always found that uniformity in step height is important. Your feet get the message immediately when they come to an odd step—a ½-in. difference is all it takes. When building stairs, make them uniform. The accident you avoid might be your own.

It isn't very difficult to build simple straight-flight stairs: All you need to know is the rise, the run, the height of the risers, the width of the treads and the amount of headroom. Transfer this information to the wood, cut and nail.

Stairwell framing was discussed on p. 49. A stairwell is the hole in the floor through which the stairs pass. It must be at least as wide as the stairs and generally a minimum of 120 in. to 130 in. long to meet code requirements for headroom. Headroom is the vertical measurement from an imaginary line connecting the front edges of all treads to the ceiling overhead. It usually must be at least 6 ft. 8 in., although 7 ft. is more comfortable for taller people.

Stairs need landings, that is, level places to plant your feet at the top and bottom, with enough room to open a door. If the stairway is 36 in. wide, the landing needs to be at least 36 in. square.

The distance that stairs travel vertically from one floor to the next (finish floor to finish floor) is called the total rise. You must know the total rise before you can figure out the height of each step (the unit rise). Measure from the rough upper floor to the rough lower floor, then check the plans to find out what kind of surface the finished floors will have. If they're both going to have the same surface, say ¾-in. hardwood flooring, then the measurement from rough floor to rough floor will be the total rise. However, if the upper floor is going to receive ¾-in. hardwood and the lower floor ⅛-in. vinyl, you need to adjust the total rise to account for the difference in finished floors. (In this example, you would need to add ⅝ in. to the rough-floor measurement.) A standard house with 92¼-in. studs, three plates, 2x10 or 2x12 joists and a ¾-in. subfloor will have a total rise of either 107 in. or 109 in.

Rise and Run

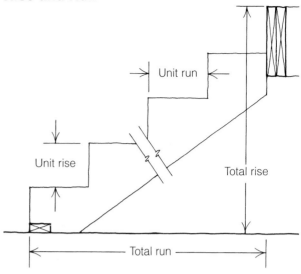

The total run is the total horizontal distance that stairs cover. Average stairs travel about 12 ft. on the level, and they take quite a bit of room out of a floor. Sometimes the total run is indicated on the plans, but often it won't be known until the stair-builder figures it out.

A comfortable step up (riser) for most adults is around 7 in. This, coupled with a unit run (tread) of not less than 10 in. (some codes require a minimum of 11 in.), constitutes the average stair of today. If the risers are too steep you may feel like a mountain climber. If the risers are too low, you'll probably find yourself taking them two at a time. Always check your code for local regulations on riser height and tread width before building stairs.

The treads of stairs sit on diagonal stringers (or carriages). The number of stringers needed to carry the stairs in a house depends on the width of the stairs. For normal stairs, 36 in. to 42 in. wide, three stringers will be required. Some codes allow stairs to be as little as 30 in. wide, but that's pretty narrow. With a 36-in. width, two people can pass with relative ease. The width shown on the plans is usually for the finished stairway. The stairs must be framed wider than the finished dimension to account for drywall on one or both sides.

Stair-Framing Calculations

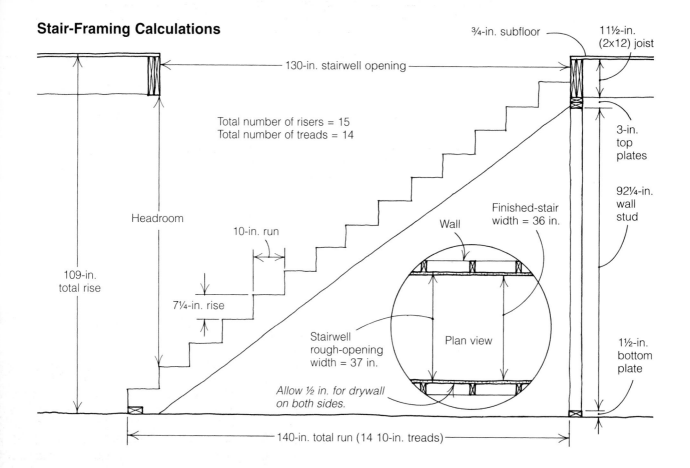

¾-in. subfloor

11½-in. (2x12) joist

130-in. stairwell opening

Total number of risers = 15
Total number of treads = 14

3-in. top plates

92¼-in. wall stud

Headroom

10-in. run

Finished-stair width = 36 in.

Wall

109-in. total rise

7¼-in. rise

Stairwell rough-opening width = 37 in.

Plan view

1½-in. bottom plate

Allow ½ in. for drywall on both sides.

140-in. total run (14 10-in. treads)

Calculating risers and treads

To calculate the number of risers needed, take the total rise (let's say it's 109 in.) and divide by 7 in., which is a good target for the unit rise. The result is 15.57. Drop the decimal fraction and you are left with 15, which is the total number of risers needed on this stair. Next, divide 109 by 15, and the answer (7.26, or 7¼ in.) is the unit rise.

This same set of stairs could be built with 14 risers instead of 15, in which case the unit rise would be about 7¾ in., which is a bit steep. With 16 risers, each step would be a little more than 6¾ in. high, which is acceptable. As a general rule, keep the unit rise around 7 in.

Once you know the number and height of risers, you must determine the width of the treads (i.e., the unit run). On many production jobs almost all treads are 10 in. wide. This standardization helps to streamline and simplify the building process. A 10-in. tread works fine with 7-in. risers. Stairs always have one tread fewer than the number of risers. The

last tread, technically speaking, is the landing and is not figured in the total run. Thus, on our stairs with a total rise of 109 in., with 15 risers of 7¼ in. each, there will be 14 10-in. treads. The total run is 14 x 10, or 140 in.

To calculate the width of individual treads from the plans, just divide the total run by the number of treads. For example, if the plans show that the total run is 144 in. and the stair has 14 treads, divide 144 by 14 for a tread width of 10¼ in.

Stringer layout

Stringer length can be calculated the same way as rafter length or by using the Pythagorean theorem (see p. 29), but for most purposes 16-ft. to 18-ft. 2x12s do the job. They will be a bit long for average stairs, but the ends of 2x12s are often split and need to be cut back to solid wood. Douglas fir works well because it is structurally very strong, but other species can also be used. Try to use straight stock without large knots, and place it on sawhorses for

Laying Out Stringers

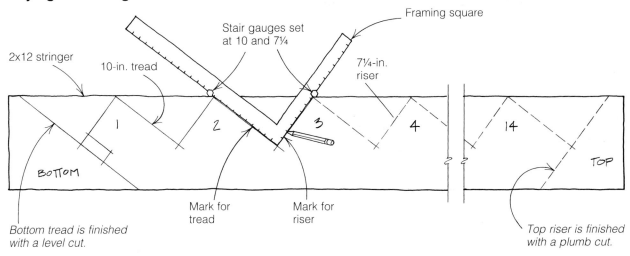

- Framing square
- Stair gauges set at 10 and 7¼
- 2x12 stringer
- 10-in. tread
- 7¼-in. riser

1 2 3 4 14

BOTTOM

TOP

Mark for tread

Mark for riser

Bottom tread is finished with a level cut.

Top riser is finished with a plumb cut.

layout. Once one stringer has been cut, it can be used as a template to mark the others.

The quickest way to lay out stringers is to use a framing square fitted with a pair of stair gauges or buttons, as shown in the photo at right. For stairs with 7¼-in. risers and 10-in. treads, place one gauge at the 7¼-in. mark on the tongue (i.e., the narrow part) of the framing square and the other gauge at 10 in. on the blade (i.e., the wider part). Begin the layout at the bottom of the stairs, with the blade toward the bottom end of the stringer. Mark across the top of the square with a sharp pencil and label this riser "1"—numbering as you go is a time-saver.

The actual number of risers that needs to laid out depends on the manner in which the stringers are attached to the landing (see pp. 195-196). The most common method is to frame the landing full-size and then hang the stairs from it, one riser down. In this case, you need to lay out only 14 risers, because the last one is the face of the landing itself. Once the necessary number of risers has been marked on the stringer, finish with a level mark at the bottom of the lowest riser and a plumb mark at the end of the highest tread.

Stringers can be laid out quickly and accurately using a framing square fitted with stair gauges.

Dropping the Stringer

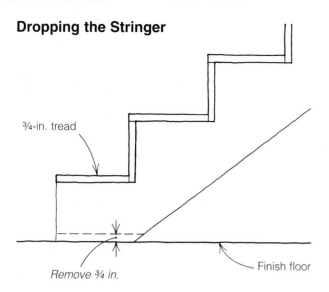

¾-in. tread

Remove ¾ in.

Finish floor

With the stringer laid out, remove from the bottom an amount equal to the width of the tread. This will bring each tread to the correct height.

Some carpenters like to back-cut the riser to make a deeper tread and give a different look to the finished stair.

Dropping the stringer

You now need to make an adjustment at the bottom of the stringer to keep the bottom riser at the same height as the others. This is called dropping the stringer, and the rule is to subtract from the bottom the thickness of the finished tread. Frequently this is ¾-in. plywood; if so, subtract ¾ in. If tread sheathing is 2x stock, subtract 1½ in. But note also what finish material will be going on the floor at the base of the stairs; if it's the same thickness as the tread sheathing, for example, it's not necessary to drop the stringer.

Once the stringer is set in place, the tread is a level cut and the riser plumb. They meet at the outside point, or nose, and form a 90° angle. Some builders vary this by giving the risers a backcut of ¾ in. or so, mainly for aesthetic reasons, but also to widen each tread (see the photo at left). The risers can be back-cut by slipping the riser gauge down the blade until the tongue of the square rests ¾ in. in from where the riser mark meets the tread mark. Then remark all the risers, making them slant back. The last tread needs to be cut ¾ in. wider to allow for a riser board to be nailed against the landing header joist (see p. 199).

To help secure the bottom of the stair, lay out a notch for a 2x4 on the bottom front of the first riser. Just take a scrap of 2x4, hold it flush with the outside corner of the first riser and scribe around it. This notch will rest on a 2x kicker that is secured to the floor.

Cutting stringers

When cutting the tread and riser lines on a stringer it is more efficient to make all the cuts in one direction before turning and cutting back the other way. Start at the bottom and cut the treads, then turn around and cut the risers. There is no need to finish the cut with a handsaw unless the stringer will be exposed, because most codes require that only 3½ in. of solid wood remain between the back of the cut and the back edge of the stringer. On a 2x12 with a 10-in. tread and 7¼-in. riser, you will have more than that even if you overcut enough for the piece to fall out.

Use a circular saw to cut out the tread and riser lines on the stringers.

Once the first stringer is cut, it can be used as a template to mark the others.

Once you've cut the first stringer, check to see that it has the proper number of risers by putting it in place in the stairwell. Nail a temporary 1x fence to the back edge of the stringer so you can use it as a template to mark the others, as shown in the photo at right. Place the first stringer flat on the second, pull it forward and snug to the fence, and mark the second stringer. If you have access to a chainsaw-type beam cutter (see pp. 145-146), you can use it to gang-cut the stringers.

Installing stringers

To hang the stringers one riser down, begin by measuring down from the top edge of the landing joists and striking a line on the face of the header where the stringer will land. This line will be down one riser height, 7¼ in., plus or minus any necessary compensation for differences in finished floor and tread surfaces.

There are several ways to fasten stringers to the header joist, as shown in the drawing on p. 196. One method is to use metal straps (18 in. to 24 in. long), which are nailed along the back edge of the stringer at the top and then bent around the stringer so the upper end can be nailed to the joist (see drawing A and the photo on p. 196). This method requires 2x6 blocking between the stringers at the top to hold them plumb. For 36-in. wide stairs with three stringers, you will need two 15¼-in. blocks. Nail a block through the side of the first stringer, flush with the top. Pull it up to the line on the header joist and secure it with nails through the block into

Stair stringers need to be fastened securely to the header joist.

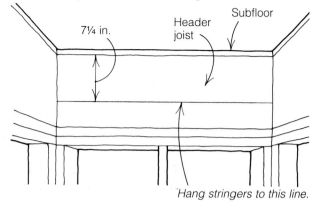

Three Ways to Install Stringers

7¼ in.

Header joist

Subfloor

Hang stringers to this line.

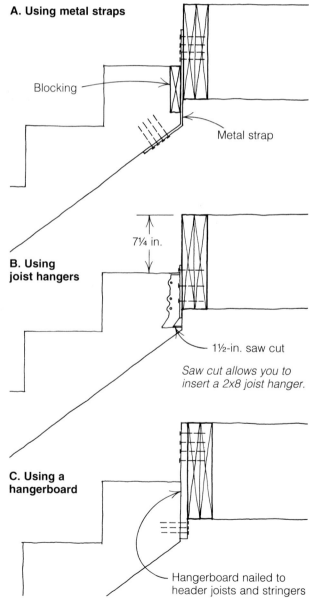

A. Using metal straps

Blocking

Metal strap

B. Using joist hangers

7¼ in.

1½-in. saw cut

Saw cut allows you to insert a 2x8 joist hanger.

C. Using a hangerboard

Hangerboard nailed to header joists and stringers

the joist. Then nail the strap to the joist. Pull up the second stringer and nail it to the first block and to the joist. Finish by nailing in the second block and the third stringer.

Another quick method of installing stringers is with 2x8 or 2x10 joist hangers (drawing B). Just make a horizontal cut about 1½ in. deep on the back of the stringer to house the bottom of the hanger. Nail the hanger to the stringer, place the stringer on the line below the landing and nail the hanger into the header joists.

A third method (drawing C) is to use a hanger-board, which is a piece of ⅝-in. or ¾-in. plywood about 15 in. wide by the width of the stairs (36 in. in our example). Nail the hangerboard to the face of the joists, flush with the landing. It should hang down no lower than the bottom of the stringer, so that it won't interfere with drywall on the under-side of the stairs. Then draw a line across the hanger-board one riser height below the landing. Secure the stringers with nails driven through the back of the hangerboard.

A 2x4 kicker at the bottom of the stairs helps to hold the stringers in place.

With the stringers nailed on at the top, cut a 2x4 kicker for the bottom the width of the stairs. Slip it into the notches on the front end of the stringers and fasten it to the floor. Then toenail the stringers to the kicker.

Drywall and skirtboards

If one or both sides of the stairs is a wall, you need to make some adjustments before hanging the stringers. By holding the stringer 1½ in. from the wall framing, you will leave plenty of room for ½-in. drywall and a 1x skirtboard (or "finish stringer") to slip in between. If the space under the stairs needs to be drywalled, you can slip the drywall all the way through to the bottom plate.

The skirtboard serves the same function as baseboard: It protects the drywall and covers the joint where stairs and wall meet. The skirtboard can be built like a reverse stringer that can fit down on top of the stairs. But it is much quicker to hold the first stringer off the wall enough to let the board slip down between it and the drywall. Hold the skirtboard up above the nose about 3 in. to 4 in. and measure its length down the stringer. Make a plumb cut at the top of the 1x skirtboard and a level cut at the bottom, and then slip it down alongside the stringer after the drywall has been installed. Any remaining

Adjusting for Drywall and Skirtboard

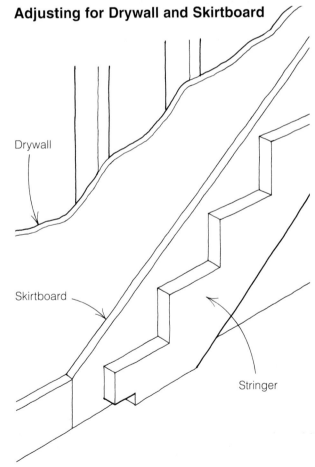

Drywall

Skirtboard

Stringer

Installing the stringer 1½ in. from the rough wall leaves room for drywall and a 1x skirtboard to slip in behind.

Pony Wall

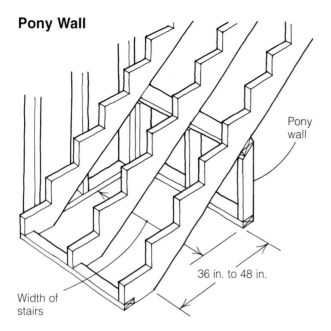

Pony wall

36 in. to 48 in.

Width of stairs

Installing Treads and Risers

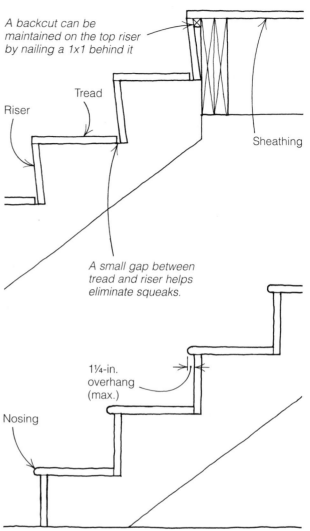

A backcut can be maintained on the top riser by nailing a 1x1 behind it

Riser

Tread

Sheathing

A small gap between tread and riser helps eliminate squeaks.

1¼-in. overhang (max.)

Nosing

gap between the stairs and the drywall or skirtboard will be covered with carpet or finish treads and risers later on.

It's a good idea to build a pony wall between the walls that enclose most staircases. A pony wall is a short wall that closes off the inaccessible area at the bottom of the stairs and helps to support the stringers at midspan (see the drawing above). Measure back about 3 ft. or 4 ft. from where the stairs and floor meet and build the pony wall under the stringers, again holding it away from the walls so that the drywall can slip past.

Cutting and installing risers and treads

If the stairs are going to be covered with carpet, they can be sheathed with material left over from the floors (at least ⅝-in. stock for risers and ¾-in. stock for treads). Rip the stock along the long edge on a table saw, or use your circular saw (carefully following a chalkline or using a guide). Both the risers and the treads will be easier to install if they are ripped slightly narrower than the actual rise or run (⅛ in. or so). Rough risers and treads don't need to fit together perfectly, especially on the backside. In fact, one potential source of stair squeaks can be eliminated if the treads and risers are prevented from touching and rubbing against each other.

Normally the treads and risers will be the same width as the stairs, but if one side of the stairway is going to be left open you may want to let them hang over so the drywall can butt up under them. Nosing, a part of the tread that protrudes beyond the face of the riser, is generally not used when stairs are to be carpeted, but if it is desired, rip the treads accordingly. Most codes allow treads to hang over up to 1¼ in.

Hardwood boards can be cut and secured directly to the stringers or to the rough-sheathed risers and treads. Many builders like to rough-sheathe their stairs and then apply the finish boards when the house is almost complete.

Construction adhesive helps to prevent squeaks and makes for a more secure stair.

Once the treads have been glued down, secure them to the stringers with 8d nails.

Treads and risers can be nailed to the stringers by hand, but it is much easier and faster to use a pneumatic nailer. The risers are nailed on first, and the treads sit on top of them. The first riser usually has to be ripped ¾ in. narrower than the rest to compensate for the dropped stringers (see p. 194). Starting at the bottom, drive two 8d nails through the riser board into each stringer. Put the next riser in halfway up the stringer. If there is any crown in the stringer, this middle riser will help straighten it out. Then go back to the bottom and begin working your way up, nailing on all the risers before beginning with the treads.

If the stringers have been given a ¾-in. backcut, nail a 1x1 strip under the floor sheathing on the square face of the landing to maintain the backcut angle for the top riser (see the top drawing at right on the facing page). In this case the last tread on the stringer needs to be cut ¾ in. wider so that all the treads will be equal when the ¾-in. riser board is nailed on. Note also that the floor sheathing must hang over enough to cover both the 1x1 and the edge of the riser board.

It is a good idea to use construction adhesive on the tread boards to help prevent squeaks from developing as the wood dries out and the nails loosen up. For extra insurance, secure the treads with drywall screws. Start at the bottom, run a good bead of adhesive on the stringers and on the riser edge, and secure each board as you go with three 8d nails or screws per stringer. When the last tread is on, jump up and down a time or two to make sure that the stair feels solid, with no bounce. Now you have easy, and safe, access to the next floor.

STAIRS WITH A LANDING

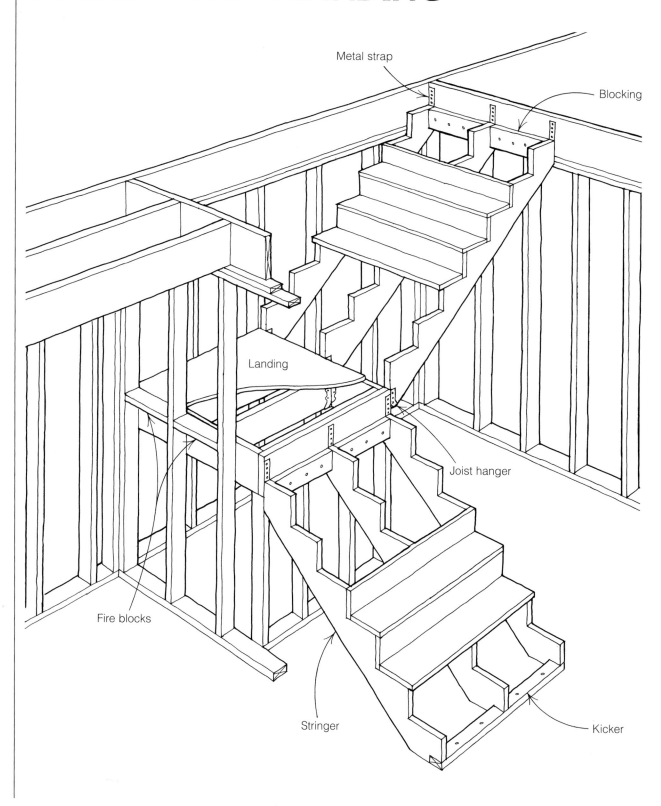

Metal strap

Blocking

Landing

Joist hanger

Fire blocks

Stringer

Kicker

Almost as common as straight-flight stairs are those that go up partway to a landing (or platform), then turn and continue. L-shaped stairs turn 90°, whereas U-shaped ones turn 180°. The plans will indicate where the stairs begin and end, the size of the landings and their location.

Before any stair stringers can be attached, the landing in the middle of the flight has to be framed. To build the landing you need to know its size, height and horizontal distance from the upper-floor landing. The size should be on the plans; on an L-shaped stair, the width and depth is normally the same as the width of the stair itself. For example, a 3-ft. wide stair will have at least a 3-ft. by 3-ft. landing. A U-shaped stair will normally have a landing that is the same width but twice as long, in this case 3 ft. by 6 ft.

The height of the landing from the bottom floor is determined by multiplying the number of risers by the height of one riser. If, for example, the plans call for seven risers to the landing and you have determined that each riser is to be 7¼ in. high, the height of the landing will be 7 x 7¼ in., or 50¾ in. Remember that this is the total height from finished floor to finished floor. Any discrepancy in floor finish thickness must be factored in as described on p. 191.

The horizontal distance between the two landings is determined by much the same method. Multiply the number of treads by the width of one tread, keeping in mind that there is one less tread than risers. So, if seven treads are needed and each one is 10 in. wide, the landing will be 70 in. from the upper landing measured horizontally. The upper stringers will be cut with a rise of 7¼ in., and a tread of 10 in. will fit between the two landings.

Landings are joisted much like a floor, using 2x8 or larger stock to make them sturdy and provide good support for stringers. Stairwells for L-shaped and U-shaped stairs are often encased by walls, and the landing joists can be nailed directly to the walls (see the drawings on p. 202). Snap a chalkline on the studs at landing height, minus the thickness of the sheathing or finish-tread material, then nail a joist to the line against the back wall. At right angles to this joist, on the side walls, nail a header joist at each end, 3 in. shorter than the width of the land-

Stairs with Landings

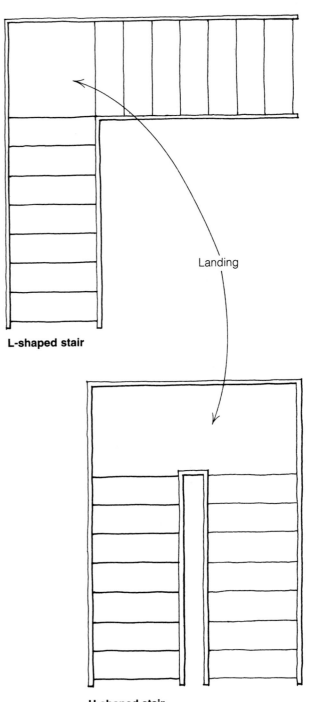

Landing

L-shaped stair

U-shaped stair

Framing the Landings

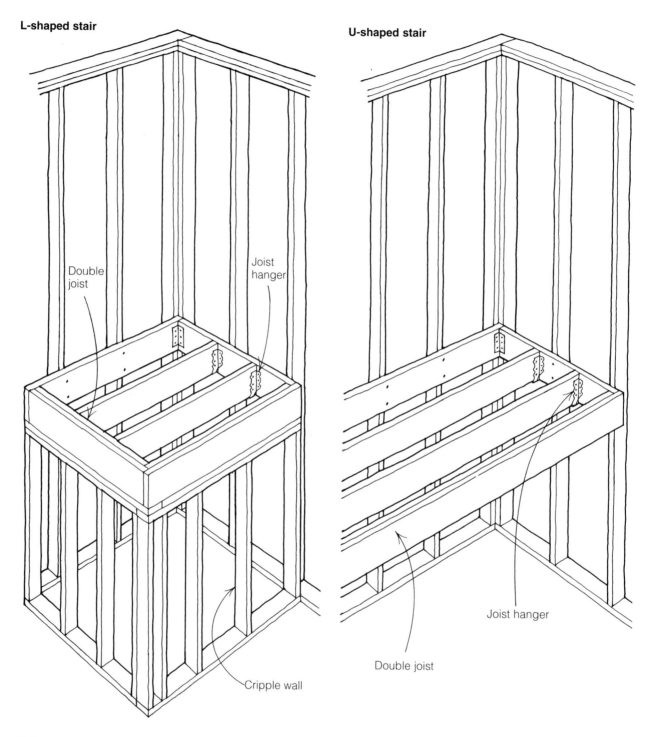

L-shaped stair

Double joist

Joist hanger

Cripple wall

U-shaped stair

Joist hanger

Double joist

This L-shaped stair landing is supported on two sides by walls. The other two sides are supported by cripple walls built under the landing.

The U-shaped stair landing joists run from wall to wall and are nailed into the studs.

ing. Hang joists on metal hangers at 16 in. o.c. between these headers, and nail a double joist to their ends to bring the landing out to the required width. The stringers will be attached to this double joist. When the stairwell is not completely enclosed by walls, nail joists where possible and then build short cripple walls to support the other sides. A 2x fire stop may be required by code in each stud space where the joists and stringers are nailed to the walls.

The stringers from the bottom floor to the middle landing are cut and installed just as they were for a regular straight-flight stairs (see pp. 194-197). Following our example, you would cut six risers at 7¼ in. and hang them from the landing. The step up to the landing constitutes the seventh riser.

The upper stringers are cut and secured a little differently. Since a total of eight risers is required, you need to cut only seven — the upper header joist will again serve as the eighth. At the bottom, however, instead of cutting the last riser square at 7¼ in. as on the lower set, continue to cut plumb down the 2x12. Then, at the tip, cut back level to leave a 1½-in. base, which will allow the stringer to sit in a joist hanger below the level of the landing (see the drawing at right).

Allowing enough room between wall and stringer for drywall and skirtboard, hold the first riser of the first stringer 7¼ in. above the landing joists, and secure the stringer to the landing with a 2x6 joist hanger. Attach the stringer at the upper landing as described on pp. 195-196. Hang the remaining stringers the same way. An alternative method is to make the middle landing about 1 ft. wider and then cut the upper stringers to rest directly on it.

With the framing complete, you can proceed to finish the risers and treads as described on pp. 198-199.

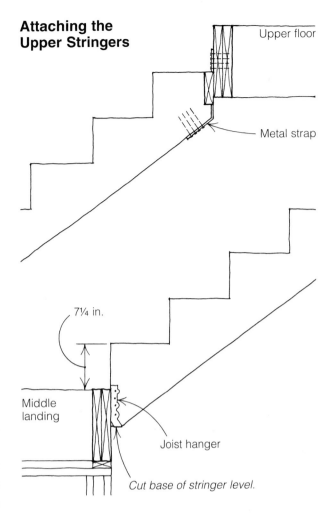

Attaching the Upper Stringers

Upper floor

Metal strap

7¼ in.

Middle landing

Joist hanger

Cut base of stringer level.

STAIRS WITH A WINDER

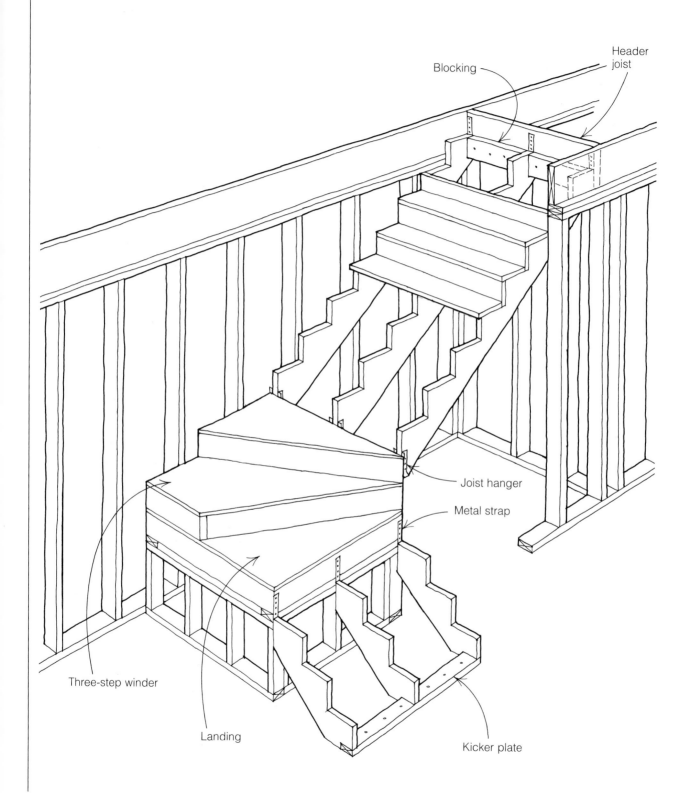

Blocking

Header joist

Joist hanger

Metal strap

Three-step winder

Landing

Kicker plate

A nother way to build an L-shaped stair is to use a three-step winder. By using a series of wedge-shaped steps rather than a landing, a winder can shorten the total run and add architectural interest to a staircase.

Because the width of the treads in a winder varies, this type of stair presents special safety concerns. Building codes regulate the shape of winder treads, but the codes vary from region to region. Some allow the treads to come to a point at one end, but most require the tread to be 6 in. wide at the narrow end and/or have a 9-in. to 10-in. wide tread at the "line of travel," which is the path a person would likely follow when ascending or descending the stairs. The line of travel is generally 12 in. to 14 in. away from the narrowest part of the stairs.

When building a winder, begin the process the same way you do with any set of stairs. First, determine the total rise and calculate the number of risers needed (see p. 192). Here, we'll follow the same example used in the section on straight-flight stairs —15 risers of 7¼ in. each.

Laying out the winder

Traditionally, carpenters have constructed winders by cutting out stringers, somewhat like those used when building a regular set of stairs. This method is familiar across the country, and it works just fine. But production framers have developed a method that eliminates winder stringers altogether; they simply build three boxes and stack them on top of each other.

For 3-ft. wide stairs, begin with a 36-in. square of ¾-in. plywood, which would normally form the landing or platform of an L-shaped stair. Divide the square into treads so that the line of travel for each tread is the same (see the top drawing at right). You can make the division using some simple math. First, snap a chalkline across the plywood from corner to corner. Then multiply the width of the stair by .52 (36 x .52 = 18.7, or about 18¾ in.). Measure 18¾ in. along the chalkline from each corner and make a mark. Then snap a chalkline from the inside corner through each of these marks. The two lines define the three treads of the winder. You can use this formula to figure out the winder layout of any

Laying out the Winder Treads

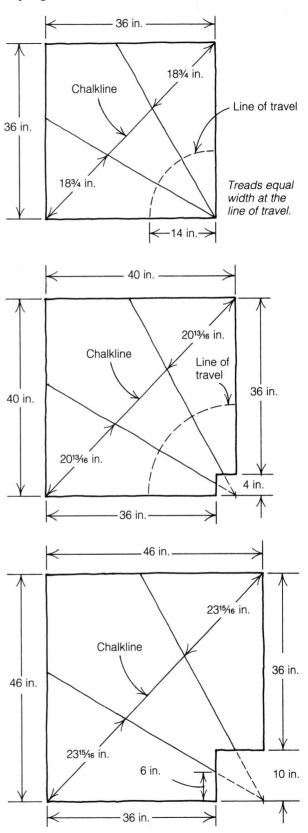

size width of stair. For example, for a 30-in. wide stair, you would measure in from each corner 15⅝ in. (30 x .52 = 15.6, or about 15⅝ in.).

If the code calls for a winder tread with a width of at least 9 in. or 10 in. at the line of travel, you'll need a 40-in. square piece of plywood to make the landing for a 36-in. wide flight of stairs. Lay out the stairs by calculating 40 in. x .52, then mark and snap lines on the plywood to lay out the treads. Next, cut out a 4-in. square at the narrow point. This reduces the stair size to 36 in. and leaves sufficient tread at the line of travel.

If you need to build stairs with a tread that is at least 6 in. wide at the narrow end, work with a plywood square that is 10 in. larger than the actual width of the stairs. So, for 36-in. wide stairs, use a 46-in. plywood square. Snap a chalkline from corner to corner, measure in 23¹⁵⁄₁₆ in. (46 x .52 = 23.92) and snap chalklines indicating the risers. Next cut out a 10-in. square at the narrow point. This reduces the stair size to 36 in. and leaves you with treads that are about 6 in. wide at the narrowest point.

Constructing the risers

After laying out the treads on the plywood square, cut one of the triangles off with a circular saw (this triangle will form the top step of the winder). The rise of each step in our example is 7¼ in., so take some 2x8 stock and rip it down to 6½ in. Once sheathed with ¾-in. plywood it will have a 7¼-in. rise. Rip enough for all of the risers. This stock will form the actual riser of each step, as well as the "joists" that support successive steps.

To make the first winder box (the landing), cut the ripped stock and build a 36-in. square frame. Sheathe it with ¾-in. plywood and place it in the stairwell at the proper height and distance from the upper landing. Next, build a second joist frame and toenail it down on top of the landing. Sheathe this second winder box with the trapezoid cut from the plywood square. Then frame, sheathe and install the third box. That's all there is to it. The landing is now a three-step winder.

Stacking the Winder Boxes

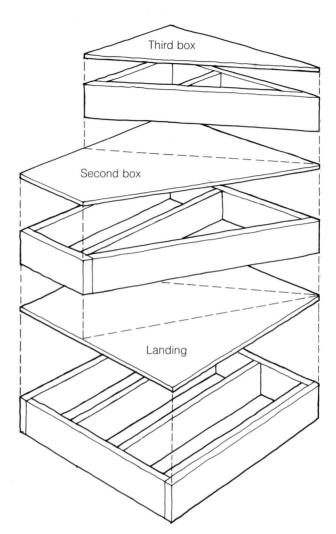

Once the landing has been framed, the second box is toenailed to it, then the third box is toenailed on top of the second.

The upper stringers are attached to the top winder with joist hangers.

Installing the stringers

Now that the winder is complete, the straight-flight stringers can be attached. If, for example, the winder landing was built 29 in. from the floor, the stair stringers from the first floor to the landing will have three 7¼-in. risers. Cut the stringers from 2x12s. They will be hung from the landing to make the fourth riser, as was done with the straight flight of stairs.

Cut the upper stringer square at the bottom, so it can slip into the joist hanger.

The stringers going up to the second floor are hung on the backside of the top winder box, just as on the landing of an L-shaped or U-shaped stair (see p. 203). Cut 1½ in. off the bottom tip of each stringer, so it will bear firmly on a joist hanger. After cutting out a stringer, it's always a good idea to check to see if it fits between the winder and the second floor. At times the treads may need to be a bit shorter or longer to run the exact distance between the two landings.

INDEX

Editor: *Jeff Beneke*
Designer/Layout Artist: *Jodie Delohery*
Illustrator: *Vince Babak*
Photographer, except where noted: *Roger Turk*
Copy/Production Editor: *Peter Chapman*
Art Assistant: *Iliana Koehler*

_____

Typeface: *ITC Stone Serif*